1st International Online Conference on Nanoscience & Nanotechnology (N@NO'2017)

Journal of Physics: Conference Series Volume 984

Johannesburg, South Africa
2 – 3 December 2017

ISBN: 978-1-5108-6318-7
ISSN: 1742-6588

Printed from e-media with permission by:

Curran Associates, Inc.
57 Morehouse Lane
Red Hook, NY 12571

Printed by Curran Associates, Inc. (2018)

Institute of Physics
Dirac House, Temple Back
Bristol BS1 6BE UK

Phone: 44 1 17 929 7481
Fax: 44 1 17 920 0979

techtracking@iop.org

Additional copies of this publication are available from:

Curran Associates, Inc.
57 Morehouse Lane
Red Hook, NY 12571 USA
Phone: 845-758-0400
Fax: 845-758-2633
Email: curran@proceedings.com
Web: www.proceedings.com

Table of contents

Volume 984

The 1st International Online Conference on Nanoscience & Nanotechnology N@NO'2017

2–3 December 2017, Johannesburg, South Africa

Accepted papers received: 28 February 2018
Published online: 16 May 2018

Preface

Papers

Theoretical and Computing Nanoscience

Graphene, Carbon and 2D nanomaterials and nanocomposites

IOP Conf. Series: Journal of Physics: Conf. Series **984** (2018) 011001 doi:10.1088/1742-6596/984/1/011001

Organizers, Sponsors and Supporters

Africa Graphene Center

ScieTechs

Department of Physics
UNISA
university of south africa

IOP Conf. Series: Journal of Physics: Conf. Series **984** (2018) 011001 doi:10.1088/1742-6596/984/1/011001

Chair of the conference

Mohammed Khenfouch, Phd. (Pr.Phys.)

Advisory Committee

Professor Teresa Cuberes (UCLM, Spain)
Professor Victor Neto (Aveiro, Portugal)
Professor Ananthakrishnan Srinivasan (IIT Guwahati, India)
Professor Vijaya Vallabhapurapu (UNISA, South Africa)
Professor Malik Maaza (Ithemba Labs, South Africa)
Professor Edward Nxumalo (SANI & UNISA, South Africa)
Professor Bakang Mothudi (UNISA, South Africa)

Organizing Committee

Mohammed Khenfouch, Phd. (Pr.Phys.)
Professor Mikael Syväjärvi (Linkoping, Sweden)
Professor Johann Boucle (Limoges, France)
Professor Izeddine (USMBA, Morocco)
Doctor Moloi Sabata (UNISA, South Africa)

Management & Administration

Doctor Guy L. Kabongo (Wits, South Africa)
Doctor Issam Derkaoui (USMBA, Morocco)

Preface

This publication forms the Proceedings of 1st International Online Conference on Nanoscience and Nanotechnology (N@NO'2017) held during 02-03 December, 2017, at Johannesburg, South Africa.

The N@NO'1 was organized by the Africa Graphene Center (AGC) in collaboration with ScieTechs, and supported by the Department of Physics of the University of South Africa (UNISA).

The conference shows two Plenary talks, two invited talks, nine Oral presentations and eight Poster presentations.

The main theme of the Conference was Nano for Sustainable Development.

The papers in this publication are grouped according to their fields of specialization.

The organization of the Congress was managed by an Organizing Committee guided by a Conference Advisory Committee. Multitasks were conducted by the management and administration members. Special thanks are due to all of them.

The Paper Reviewers and Co-Reviewers went through several abstracts and full length papers and finaly accepted 23 abstracts and 16 paper. Their untiring efforts are gratefully acknowledged.

The contributions made by the keynote and invited speakers, the presenters and the authors of papers as gratefully acknowledged.

The Conference proceedings have been published by IOP journal of Physics: conference series. Their efficient and high-quality work is appreciated.

N@NO'2017 team

IOP Conf. Series: Journal of Physics: Conf. Series 984 (2018) 011001 doi:10.1088/1742-6596/984/1/0

N@NO 2017

About N@NO'2017

N@NO'2017, also named (N@NO'1) is a top-level international conference offering a virtual platform of knowledge for scientists and engineers from around the world to share experience and contribute in the advancement of nanoscience and nanotechnology.
The conference will provide parallel symposium in the main topics of nano where e-presentations and e-posters will be displayed.
Selected papers will be published in an ISI journal.
All presented abstracts by one of the authors will appear in the conference proceedings.
The conference language will be English.

Topics

Theoretical and Computational Nanoscience.
Graphene, Carbon and 2D nanomaterials and nanocomposites.
Synthesis and Characterization Techniques of Nanoscale Materials.
Nano-electronics, Nano-optics, Nano-optoelectronics, Nanophotonics, Nanoplasmonics.
Applied Nanoscience, Nanotechnology for Energy and Environment, Water Purification…
Biomaterials and Nanobiotechnology.

Peer review statement

All papers published in this volume of *Journal of Physics: Conference Series* have been peer reviewed through processes administered by the proceedings Editors. Reviews were conducted by expert referees to the professional and scientific standards expected of a proceedings journal published by IOP Publishing.

IOP Conf. Series: Journal of Physics: Conf. Series **984** (2018) 012001 doi:10.1088/1742-6596/984/1/012001

Polarizability and binding energy of a shallow donor in spherical quantum dot-quantum well (QD-QW)

K Rahmani[1], Y Chrafih[2], S M'Zred[3], S Janati[3], I Zorkani[3], A Jorio[3] and A Mmadi[4]

[1]LIRST Faculty of Polydisciplinary, Beni Mellal-Morocco
[2]LDD Faculty of Sciences and Technologies, Beni Mellal-Morocco
[3]LPS, Faculty of Sciences Dhar Mehraz, Fès-Morocco
[4]Faculty of Sciences and Technologies, Moroni -Comores

Email: kh.rahmani@hotmail.fr

Abstract. The polarizability and the binding energy is estimated for a shallow donor confined to move in inhomogeneous quantum dots (CdS/HgS/CdS). In this work, the Hass variational method within the effective mass approximation in used in the case of an infinitely deep well. The polarizability and the binding energy depend on the inner and the outer radius of the QDQW, also it depends strongly on the donor position. It's found that the stark effect is more important when the impurity is located at the center of the (QDQW) and becomes less important when the donor moves toward the extremities of the spherical layer. When the electric field increases, the binding energy and the polarizability decreases. Its effects is more pronounced when the impurity is placed on the center of the spherical layer and decrease when the donor move toward extremities of this spherical layer. We have demonstrated the existence of a critical value $\left(\frac{R_1}{R_2}\right)_{cri}$ which can be used to distinguish the tree dimension confinement from the spherical surface confinement and it's may be important for the nanofabrication techniques.

1. Introduction

In this last decade the advances in nanofabrication technology have made it possible to manufacture low-dimensional semiconductor materials such as quantum well (QW), quantum well wire (QWW) and finally quantum dots (QDs) structure where the electron is confined in three directions. With their reduced dimensionality, new effects can be observed and studied. In particular, the binding energies of shallow donors are expected to become important while they are known to be low in bulk materials [1-2]. The semiconductor quantum dots (QDs) structure, often called artificial atoms, have great potential for device applications in the nanometer scale [3]. Therefore, in recent years, both experimental and theoretical studies of the electronic and optical properties of homogenous quantum dots (QDs) have attracted the considerable attention of many scientists in condensed matter physics and applied sciences [4-6]. In this structure, a new type of quantum dot (QD) structure, called a quantum dot-quantum well (QDQW) or inhomogeneous quantum dots (IQD), has been fabricated and studied. Theses structures experimentally are composed of two semiconductor materials one of which, with the smaller band gap, is embedded between a core and outer shell of the material with the larger band gap.

IOP Conf. Series: Journal of Physics: Conf. Series **984** (2018) 012001 doi:10.1088/1742-6596/984/1/012001

Nowadays, a synthesized inhomogeneous spherical quantum dot with a central core and one or several layers of shells (such as CdSe/ZnS, InAs/ZnSe, CdS/PbS) called core–shell structure quantum dot or quantum dot–quantum well (QDQW) attracts many scientists' interests [7,8].
Xi Zhang and al [9] studied the third-order nonlinear susceptibilities associated with intersubband transition are theoretically calculated for ZnS/CdSe cylindrical quantum dot quantum well. X.N.Li u and D.Z.Yao [10] interested on the study for nonlinear optical properties of the CdSe/ZnS quantum dot quantum well (QDQW). Xing and al. [11] focus problem of bound polarons in quantum dot quantum well (QDQW) structures is studied theoretically.
For more details on the chemical fabrication of this new artificial structure, we refer to references: CdSe/ZnS/CdSe/ZnS [12], In(Ga)As/GaAs [13] and InAs/ZnSe [14].
Understanding the Impurity states in these structures is an important problem in semiconductors technology. Many works have been devoted to the studies of impurity states in the quantum well, quantum wire and quantum dots. Recently, J.Silva and al [15] have calculated the donor binding energy as a function of the donor position in QD with an infinite spherical well and for different radius of the structure. They found that the donor binding energy decreases when the donor position increases reaching a minimum, as the donor position is equal to the radius of the QD. Recently the dipole polarizability of the harmonic oscillator of nanodot has been calculated analytically [16-17].
Many theoretical attempts have been made to understand the polarizability [18-20].
For QDQW, to the best of our knowledge no work has been done to treat the polarisability and evaluate the effect of the donors position. The quantum confined stark effect has been the subject of numerous experimental and theoretical investigations [21] which demonstrated that the binding energy are shifted to low energies in the presence of an uniform electric field. The effect of the latter is to change the energy levels in the QD and giving rise to a polarization of the distribution of charge carriers and has other energy transition between quantum states [22].
D Schoss et all are presented a theoretical and experimental study on the inhomogeneous quantum dots (CdS/HgS/CdS) [23].
In this work, we use a variational method to calculate the polarisability and the binding energy of an electron bound to a donor impurity confined to move in inhomogeneous quantum dots (CdS/HgS/CdS) in the strong and the moderate regime of confinement. We will examine the electric field and the position of a donor impurity effects on the binding energy and the polarizability. We will examine the effect of the size of the dots.

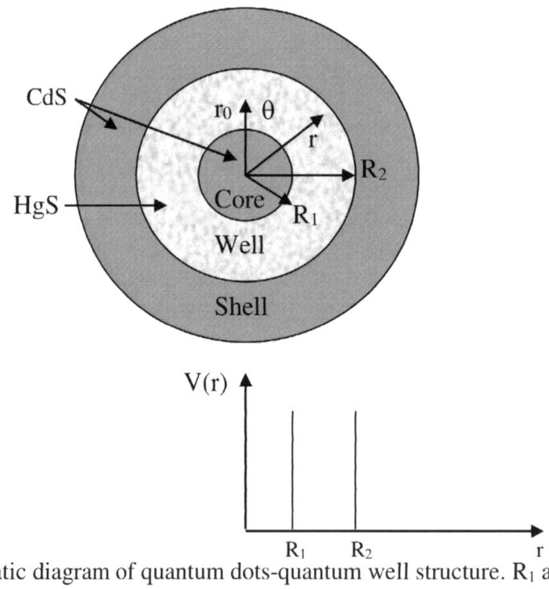

Figure 1. Schematic diagram of quantum dots-quantum well structure. R_1 and R_2 are the inner and the outer radius of the QDQW respectively.

2. General formalism

We Consider a hydrogenic impurity located at the position r_0 of a quantum dot-quantum wells (QDQW) made from [CdS (Core)/HgS(well)/CdS (Shell)] subjected to the action of an electric field outside F. The confining potential is assumed as a model of an infinite deep well. In the effective mass approximation, the Hamiltonian system is written:

$$H = H_0 + V(r) + W \tag{1}$$

Where V(r) is the confining potential. It's written by:

$$V(r) = \begin{cases} 0 & R_1 \prec r \langle R_2 \\ \infty & r \prec R_1 \quad or \quad r \rangle R_2 \end{cases} \tag{2}$$

R_1 et R_2 denotes the inside and outside radius of the quantum dot-quantum wells, respectively.
H_0 is the Hamiltonian in the absence of the electric field given by:

$$H_0 = -\frac{\hbar^2}{2m^*}\Delta - \frac{e^2}{\varepsilon_0 |\vec{r} - \vec{r}_0|} \tag{3}$$

Where $$|\vec{r} - \vec{r}_0| = \sqrt{r^2 + r_0^2 - 2 r r_0 \operatorname{Cos}\theta}$$

ε_0 is the dielectric constant and m^* is the effective mass of the electron in the QDQW.

$$W = e\ f\ (r\cos\theta - r_0) \tag{4}$$

Thus we introduce the dimensionless parameter characterizing the strength of the electric field $f = \frac{e a^* F}{R^*}$.

In a system of reduced units the Hamiltonian of an electron bound to a donor impurity in the dot ball is given by:

$$H = -\nabla^2 - \frac{2}{|\vec{r} - \vec{r}_0|} + f(r\cos\theta - r_0) + V(r) \tag{5}$$

The Schrödinger equation expressed in spherical coordinates (r,θ,φ) is written as:

$$H\psi(r,\theta,\varphi) = E\psi(r,\theta,\varphi) \tag{6}$$

We use a variational method approach to determine the ground state binding energy and the polarisability, We adopt the wave function given by :

$$\psi = \psi_0\ [1+\beta\ f\ (r\ \cos\ \theta\ -\ r_0\ )] \tag{7}$$

ψ_0 is the wave function in the absence of the electric field (f = 0) given by:

$$\psi_0 = \frac{\operatorname{Sin}[K(r-a)]}{r} e^{-\alpha|\vec{r}-\vec{r}_0|} \tag{8}$$

Where α and β are the variationals parameters (which takes into account the presence of the electric field).

The energy is obtained by the minimization with respect to the variationals parameters:

$$E_t = \min_{\alpha,\beta}\left\{\frac{\langle\psi|H|\psi\rangle}{\langle\psi|\psi\rangle}\right\} \tag{9}$$

The binding energy E_b of the Impurity is given by:

$$E_b = E_{Sub} - E_t \tag{10}$$

Where E_t and E_{Sub} represents the ground-state energy of the structure QDQW with and without the impurity respectively.
The polarizability can be calculated from the dipole moment and is defined as:

$$P = -e(\langle|\vec{r} - \vec{r}_0|\rangle_{f \neq 0} - \langle|\vec{r} - \vec{r}_0|\rangle_{f=0}) \tag{11}$$

$$P = \alpha_p F \tag{12}$$

We use the effective Bohr radius $a^* = \dfrac{\hbar^2 \varepsilon_0}{m^* e^2}$ and the effective Rydberg $R^* = \dfrac{m^* e^4}{2\hbar^2 \varepsilon_0^2}$ as the units of length and energy (respectively).

3. Results and Discussion

In this section we will discuss the results obtained by applying the direct variational method for calculating the electric polarizability by the formula dipole moment P and the binding energy of a hydrogenic impurity in the absence and presence of an electric field in the quantum dot-quantum well (QDQW). Numerical calculations are applied to a spherical inhomogeneous quantum dot [CdS (Core)/HgS (well)/CdS (Shell)]. The physicals parameters characterizing the system are given in the table 1 [24]:

Table 1. The physical parameters characterizing the system

Material	**m^*/m_0**	**ε_0**	**ε_∞**	**a* (A°)**	**R* (eV)**	**Eg (eV)**
CdS	0,2	9,1	5,5	5,818	3,65	2,5
HgS	0,036	18,2	11,36	5,851	5,0	0,5

In our calculations, two extreme cases that present themselves: $R_1 = 0$ (Homogeneous Quantum Dot "HQD" of radius R_2) and $R_1 \rightarrow R_2$ for R_2 fixed which corresponds to an infinitely thin spherical layer. We will present the results for the cases of a HQD and IQD.

3.1 Homogeneous Quantum Dot "HQD"

In Figure 2. We present the donor binding energy as a function of the donor position in a spherical quantum dot ($R_1 = 0$) and for different radius of the structure (R_2=1a*, R_2=2a*, R_2=4a*, R_2=6a* and R_2=8a*). This figure shows that the binding energy is more important in the middle $\left(\dfrac{r_0}{R_2} = 0\right)$ and decreases when the impurity moves toward the extremities of the quantum dot to reach its minimum for $\left(\dfrac{r_0}{R_2} \rightarrow 1\right)$. Thus, we see that for a fixed position of the impurity, the binding energy increases with decreasing the dot radii R_2. It is due to the distribution of the electronic wave function. Therefore, the decrease of the geometric dimensions of the quantum dot creates a compression wave function and the electron is closest to the impurity. It is causes the increased binding energy. Our results are in good agreement with those found by Silva-Valencia and all [2, 25].
In figure 3, we plotted the electric polarizability of an impurity associated with an electron according to the radius of the QD for different values of the electric field. In this figure, we note that the polarizability decreases with increasing the intensity of the electric field. Ours results show that the polarizability increases as the radius of the dot increases. We remark that the electric field effect is more pronounced for large dots. That we can explain that by for larges quantum dots, the electronic confinement is negligible and therefore the results tend to the case of semiconductor solid 3D. The intensity of the electric field increases the confinement of the electron and reduces the polarizability.
The electric polarizability of an impurity depending on the position of $\dfrac{r_0}{R_2}$ quantum dot is plotted in figure 4. In this figure, we show that the electric polarizability increases with the donor position until

it reaches the report $\frac{r_0}{R_2} \approx \frac{1}{2}$. For $\frac{r_0}{R_2} \succ \frac{1}{2}$ polarizability decreases with increasing the size of the quantum dot. This implies that the confinement effect is more significant than the effect of the electric field in the case of weak non-significant inverse positions in major positions. Its effect is more pronounced when the impurity is placed between the middle of the dot $\frac{r_0}{R_2} \approx 0{,}4$ to $\frac{r_0}{R_2} \approx 0{,}8$ and decreases as the donor moves to the ends of the dot. We also note that the electric polarizability decreases as we increase the electric field.

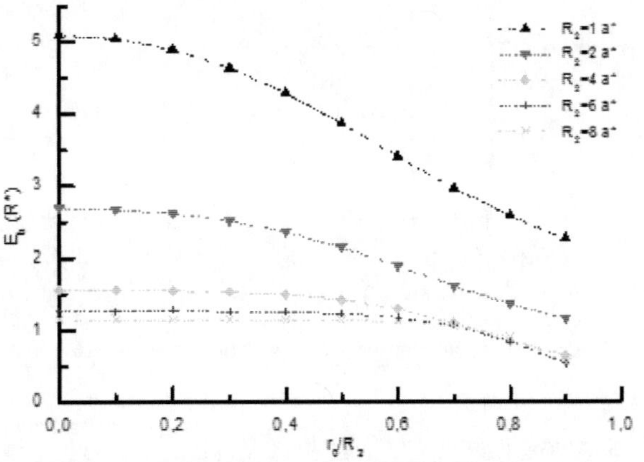

Figure 2. Variations of the donor binding energy E_b as a function of the donor position (r_0 / R_2) for different values of the dot radius R_2 = 1 a*, 2 a*, 3a*, 4 a* and 6 a*.

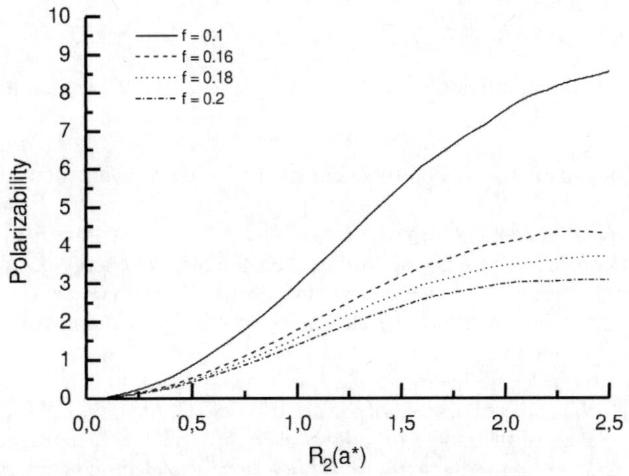

Figure 3. Polarizability of an impurity function of the radius R_2 for different values of the electric field f = 0,1, f = 0.16, f = 0 .18 and f = 0.2.

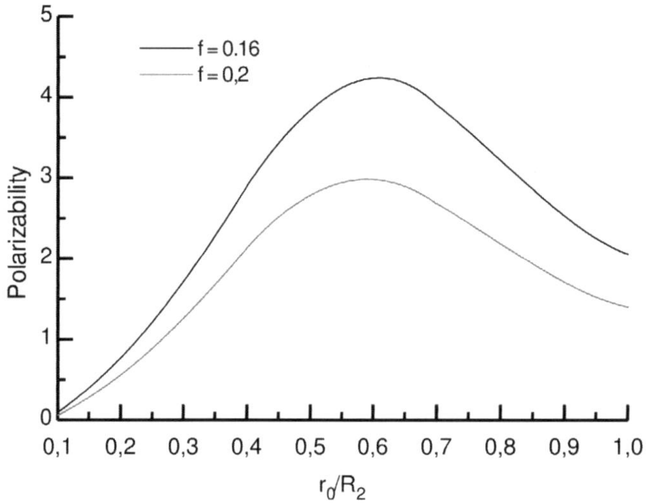

Figure 4. Polarizability of an impurity donor function deposition r_0 / R $_2$ for two values of the electric field.

3.2 Inhomogeneous Quantum Dot ''IQD''

In figure 5, we present the binding energy E_b as a function the ratio (R_1/R_2) for two different regimes of confinement has R_2 = 1a* and R_2 = 2a* and for two values of electric field f = 0,4 and 0.8. We see that the binding energy depends strongly of the ratio $\frac{R_1}{R_2}$ increasing from 0 to 1 and presents a minimum $\left(\frac{R_1}{R_2}\right)_{crit}$ witch can be used in nanofabrication of QDQW. When the inner radius R_1 a increases, the donor binding energy moves from the ''HQD'' situation $\left(\frac{R_1}{R_2} \to 0\right)$ where the kinetic energy is dominating to a small spherical layer $\left(\frac{R_1}{R_2} \to 1\right)$ where the coulomb energy prevails and the donor can be considered as a fixed.

We note that when the electric field increases, the binding energy decreases and we also note that the electric field effect becomes more pronounced when the ratio (R_1/R_2) tend to 1.

We have reported in figure 6, the electric polarizability of the donor impurity according to the ratio R_1/R_2 of the quantum dot quantum well (QDQW) for four values of outer radii R_2 and for an electric field strength (f = 0.2). This figure illustrates the competition between the geometric confinement (the size of QDQW) and containment of the electric field on the polarizability.

For larges values of the ratio R_1/R_2 (low spherical layer), the geometric confinement is important and the wave function of the electron is very localized. The electric polarizability of the impurity becomes less important and is relatively insensitive to the outer radii R_2 of the QDQW. For low values of the ratio R_1/R_2 (large spherical layer), the effect of electric field predominates. The polarizability is strongly depend of the size of quantum dot-quantum well. We note that the polarizability decreases as the R_1/R_2 ratio increases.

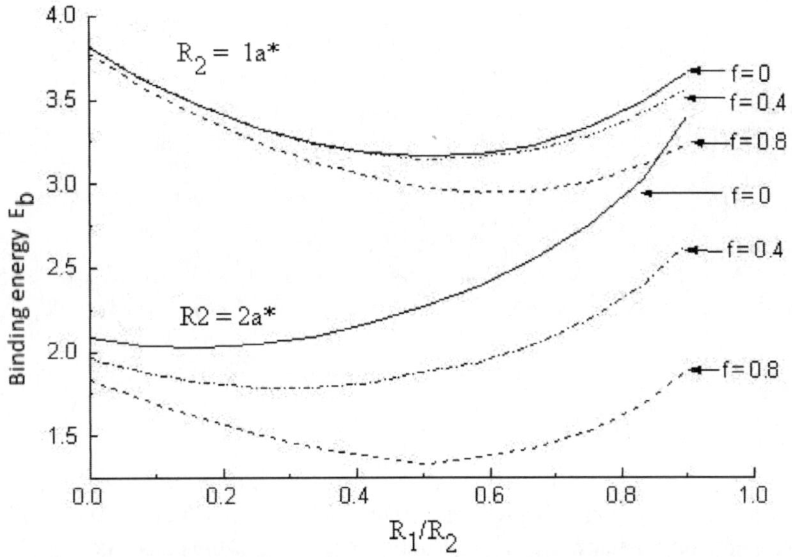

Figure 5. Variations of the binding energy E_b as a function of the radio a/b for two different values of the outer radius of the QDQW, R_2 = 1 a* and 2a* and two values of the electric field f = 0.4 and 0.8.

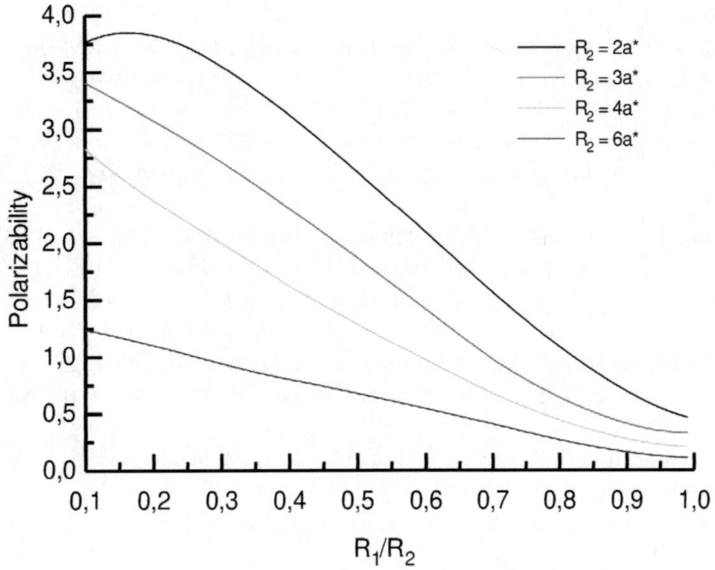

Figure 6. Polarizability of an impurity depending on the ratio $R_1/R_2$for different values of outer radii R_2.

4. Conclusion

In conclusion, we have studied the polarizability and the binding energy of an impurity located in Inhomogeneous Quantum Dots (IQD) in the presence of an electric field. It's clear that they strongly depend on the geometric confinement, the impurity position and the electric field.

We have shown the existence of a critical value $(R_1/R_2)_{crit}$ which may be important for the nanofabrication techniques.

We think that the present study will allow a better understanding of the behaviour of these new structures. We believe that our results of the polarisability in IQD will be relevant in the interpretation of experimental results. Unfortunately, we could not compare our results as no explicit experimental data are available in the literature.

References:

[1] Djillali Nasri, Nadir Sekkal, *Physica E***42** (2010) 2257–2263
[2] I. F. I. Mikhail and I. M. M. Ismail, *phys. stat. sol. (b)* **244**, No. 10, 3647–3659 (2007)
[3] M. Kırak, S.Yılmaz, M. Sahin and M. Gençaslan, *Journal of Applied Physics* 109, 09, 4309 (2011)
[4] Jayanta Ganguly, Surajit Saha, Aindrila Bera, Manas Ghosh, Superlattices and Microstructures 98 (2016) 385-399
[5] Aindrila Bera , Surajit Saha, Jayanta Ganguly, Manas Ghosh, Journal of Physics and Chemistry of Solids 98 (2016) 190–197
[6] Erfan Owji, Alireza Keshavarz, Hosein Mokhtari, Superlattices and Microstructures 98 (2016) 276-282
[7] Xiaobo Feng, Guiguang Xiong, XI Zhang, et al., *Physica B* **383** (2006) 207.
[8] Shaohua Gong, Duanzheng Yao, Xiaobo Feng, et al., *Microelectron. J*. **37** (2006) 904.
[9] Xi Zhang, Guiguang Xiong , Xiaobo Feng *Physica E* **33** (2006) 120–124
[10] X.N. Liu and D.Z. Yao, *Eur. Phys. J. B* 78, 95–102 (2010)
[11] Xing. Yan, Wang. Zhi-Pin and Wang Xu, *Chinese Physics B*, Vol 18 No 5 (2009), 1674-1056/1935-07
[12] A. Salman Ogli, A. Rostami, *J Nanopart Res* (2011) 13:1197–1205
[13] G. Sek1, J. Andrzejewski and al. *Semicond. Sci. Technol.* **24** (2009) 085011 (5pp)
[14] E. Assaid, E. Feddi, M. Khaidar and F.Dujardin, *physica scripta*. Vol. 63, 329-335 (2001)
[15] D. Nasri, N. Sekkal, *Physica E*42 (2010) 2257–2263
[16] J. Silva-valencia, N. Porras-Montenegro, IC/95/216 (1995)
[17] E. Kilicarslan, S.akiroglu, E. Kasapoglu, H. Sari, I. Sökmen, *Superlattices and Microstructures* 48 (2010) 305-311
[18] R. Khordad, A. Gharaati b, M. Haghparast, Current *Applied Physics* **10** (2010) 199–202
[19] A. John Peter, *Superlattices and Microstructures* **44** (2008) 106–111
[20] Surajit Sahaa, Jayanta Gangulyb and Manas Ghoshc, *The Open Optics Journal* (2015) 9, 7-13
[21] G. Rezaei , S. Mousavi, E. Sadeghi, *Physica B* **407** (2012) 2637–2641
[22] C.A. Duque, A. Montes, A.L. Morales, *Physica B* 302-303(2001) 84-87
[23] D Schoss, A Mews, A Eyshmüler and H Weller, Physical Review B, Vol 49 N 24 (1994) 17072-17078
[24] S Mzerd, K Rahmani, S Janati, Y Chrafih, I Zorkani And A Jorio, *IOP Conf. Series: materials Science and Engineering* **186** (2017) 012015
[25] I Zorkani, K Rahmani, M El Yahyaoui And A Didi Seddik, *Phys Low-Dim Struct,* 11/12 (2003) pp 85-100

Effect of conduction band non-parabolicity on the intersubband transitions in ZnO/$Mg_xZn_{1-x}O$ quantum well heterostructures

L Moudou[1], Y Chrafih [1], K Rahmani[2] and I Zorkani[3]

[1]LDD Faculty of Sciences and Technologies, Beni Mellal-Morocco
[2]LIRST Faculty of Polydisciplinary, Beni Mellal-Morocco
[3]LPS, Faculty of Sciences Dhar Mehraz, Fès-Morocco

Email: younescharafih@gmail.com --- kh.rahmani@hotmail.fr

Abstract. In this paper, we have calculated the electronic states and the coefficient of transmission in ZnO/$Mg_xZn_{1-x}O$ quantum well structures (QW), with 20% of Magnesium in both the parabolic and the non-parabolic cases. Our calculations are performed in the context of the approximation of the envelope function formalism, and using the finite difference method. The results show that the intersubband transition energy increases rapidly with well width until L_w=5nm and becomes almost constant (specially transitions E_{13} et E_{23}). Wavelength λ_{23} decreases with well width until L_w=5nm and becomes constant. The non-parabolicity effect is more pronounced for small QW ($L_w \leq$ 5nm) and less marked in narrow and large QW. Also, we are studied the coefficient of transmission. We notice that when the height of barrier increases the coefficient of transmission decreases. It will be necessary to provide more energy to the electron so that it can cross the barrier. We also notice the variations related to a phenomenon of reflection quantum.

1. Introduction:

Semiconductor materials II-VI with a large band gap of the ZnO/MgZnO system are gaining increasing interest; they are the potential competitors of the other large family of materials with a large band gap, the III-V nitrides. Like the nitrides, ZnO crystallizes in the wurtzite phase then the presence of the particular effects related to the polarization with internal electric fields more intense than those in the nitrides III-V, due to their applications in the field of fabrication of the photodetectors and diode lasers [1-5]. However, ZnO has several advantages, the ZnO substrate is naturally solid ZnO of good crystalline quality, it is not very expensive and easy to manufacture [6-7]. Intersubband transitions in quantum wells of ZnO/MgZnO are theoretically and experimentally determined for the purpose of improving optoelectronic properties, and making it an ideal candidate for optoelectronic devices, more precisely in the ultraviolet regions [8-11].

Recently, several groups interested on intersubband transitions in quantum wells composed of semiconductors due to their potential in ultraviolet regions. We found that Bhattacharya et al.[12] experimentally studied the transmission spectra of MgZnO thin films with different Mg to Zn by changing the thicknesses of ZnO layer, and found that the films showed high (>80%) transparency, and the absorption edge shifted as the thickness of the ZnO sublayers decreases. Ohno et al.[13-14] studied the intersubband transitions in ZnO-MgZnO and have reported that there are unique properties for intersubband absorption in ZnO quantum well. In this work, we adopt a numerical method to determine the intersubband transitions of the conduction band in ZnO/MgZnO, by solving the Schrödinger equation, under the effect of the approximation of the envelope function and considering the impact of the polarization electric fields in the quantum wells and barriers. In Part 2, we present the study of the calculations of the structure of the band and show the effects of non-parabolicity on the positions of confined levels. In Part 3, we present the study of the transmission coefficient.

2. General formalism:

Under the effect of the approximation of the envelope function, the wave functions and the energy levels are obtained by solving the Schrödinger equation:

$$\left(\frac{p^2}{2m_0}+V(r)+U_H(z)\right)\varphi(r)=E\varphi(r) \qquad (1)$$

Where V(r) is the crystalline potential and UH (z) is the potential of the heterostructure.

$$U_H(z)=\begin{cases} eF_b z - eF_w(z_2-z_1)+\Delta E_c & z < z_1 \\ eF_w(z-z_1) & z_1 \le z \le z_2 \\ eF_b(z-z_3)+\Delta E_c & z > z_2 \end{cases} \qquad (2)$$

Where z_1 and z_2 determine the barrier and well interfaces positions: $z_3 = z_2 + z_1$. The position z_1 taken as the origin of the potentials.

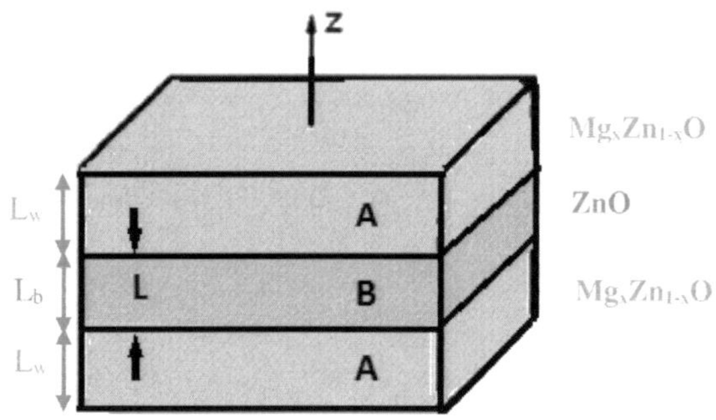

The potential of a hétérostructure $U_H(z)$ often results from the strip discontinuity (ΔE_c : offset) at the interface of the two materials due to the difference in the position of the conduction bands. Nevertheless, this time, the presence of internal electric fields (F_w and F_b) in each of the two materials will influence it.

$$F_w=\frac{(P_b-P_w)L_b}{\varepsilon_0(\varepsilon_w L_w+\varepsilon_b L_b)} \quad and \quad F_b=\frac{-(P_b-P_w)L_w}{\varepsilon_0(\varepsilon_w L_w+\varepsilon_b L_b)}=-F_w\frac{L_w}{L_b} \qquad (3)$$

Where ε_0, $\varepsilon_{w(b)}$ and P_b, P_w are respectively the free static dielectric constant, the static dielectric constant of the well (barrier) and the total polarizations in the well and in the barrier [15-16].

The maximum value of the internal electric field $F_{MAX}=\Delta P/\varepsilon_B$ therefore depends directly on the magnesium proportion in the barrier. An approximately linear increase was measured by time-resolved photoluminescence measurements [17]:

$F_{MAX}(x_{Mg})=3.95\,x_{Mg}\ MV/cm$ That to say $F_{MAX}=0.79\ MV/cm$

The effective mass, the relative permittivity are calculated from Vegard's law (linear interpolation between the values of ZnO and MgO) and the gap varies with the composition according to the quadratic law:

$$E^{g}_{Al_{1-y}In_{y}N}(y) = y.E^{g}_{InN} + (1-y).E^{g}_{AlN} - y.(1-y).b \tag{4}$$

Where the bowing parameter b = 0 eV for the MgZnO alloy where the Mg fraction is less than 36% [18]. Moreover $\Delta E_c = 0.7.\Delta E_g = 210\ meV$.

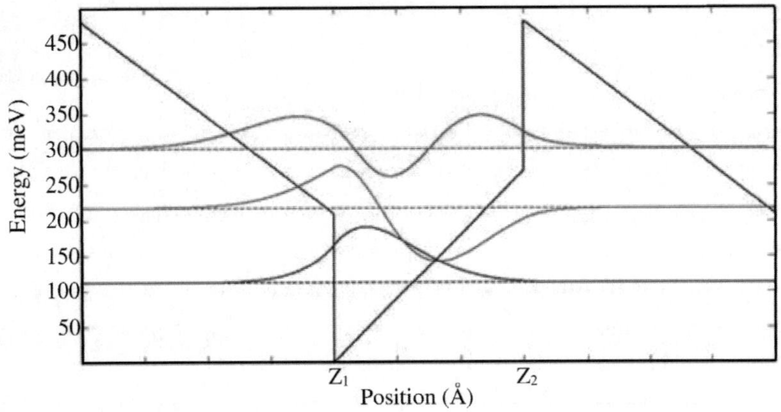

Figure 1. : Conduction band profile, confined level positions and associated envelope functions in $ZnO/Mg_{0.2}Zn_{0.8}O$ quantum well for L_w = 60 Å and L_b = 80 Å

Table1. Structural and electronic parameters of ZnO and MgO [19].

Parameter	ZnO	MgO
Lattice constant a (Å)	**3.2**	**4.2**
Gap energy (eV)	**3.37**	**5.4**
Conduction band effective masses m_e	**0.28**	--

The calculations were developed in the parabolic approximation and in the nonparabolic case where the effective mass depends on the energy according to the Nilson model [19]:

$$m_w(\varepsilon) = m_w(1+\frac{\varepsilon}{E_w}) \quad and \quad m_b(\varepsilon) = m_b(1-\frac{Vs-\varepsilon}{E_w}) \tag{5}$$

$E_{w(b)}$: The gap energy of the well (barrier)
$m_{w(b)}$: The effective mass of BC electrons in the parabolic approximation in the well (barrier)

In the framework of the envelope function model, the Schrödinger equation is reduced to the following differential equation:

$$\left(\frac{-\hbar^2}{2}\frac{d}{dz}\left(\frac{1}{m^*(z)}\frac{d}{dz}\right)+U_H(z)\right)\chi_i(z)=E_i\ \chi_i(z) \tag{6}$$

The numerical resolution is based on the finite difference method. It considers the structure as a succession of n + 1 layers of very small but equal thicknesses, Δz. In each layer, the mass (m_n), the potential (U_n) and the envelope functions (χ_n) are considered as constants. The envelope functions in the different layers in this model are linked together by:

$$\chi_{n+1} = \chi_n \left(\Delta z^2 k_n^2 + \frac{m_n}{m_{n-1}} + 1 \right) - \frac{m_n}{m_{n-1}} \chi_{n-1} \tag{7}$$

Where : $$k_n^2 = 2\frac{m_n(U_n - E)}{\hbar^2}$$

3. Result and Discussion:

3.1 Electronic states in ZnO/MgxZn1-xO quantum well

The confined levels are relatively sensitive to variation in the width of the barrier, in particular for low L_b. The positions of the fundamental level and the 1st excited level, located below V_s increase when the barrier is widened. On the contrary, those located in above of V_s descend in energy. It should be noted that those energies variations are generally low after a certain width of the barrier. Considering, as well, the value of L_b = 8 nm to study the effect of the well width. The following figure is a graphical representation of the different confined levels evolution and transitions energies in function of well width L_w:

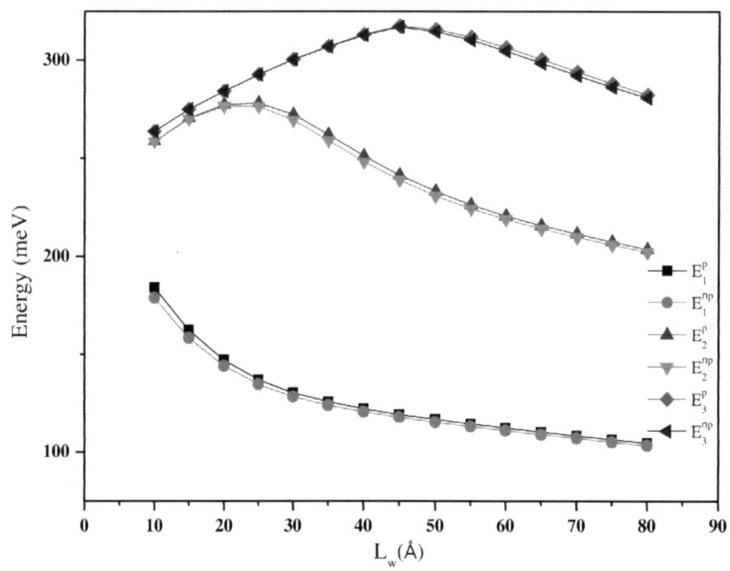

(a)

According to the curve in figure 2. (a), it can be seen that when the well width decreases each confined levels increase in energy up to a certain width, which they have a maximum value and below which their energies decrease, L_w = 45Å for E_3 And L_w = 25Å for E_2. When a confined level energy enters in above of V_s, its position undergoes a slowdown due to the combined effect of the continuum and confining field of the barrier. In addition, for broad well widths of the variation of confined levels energies are overall weak.

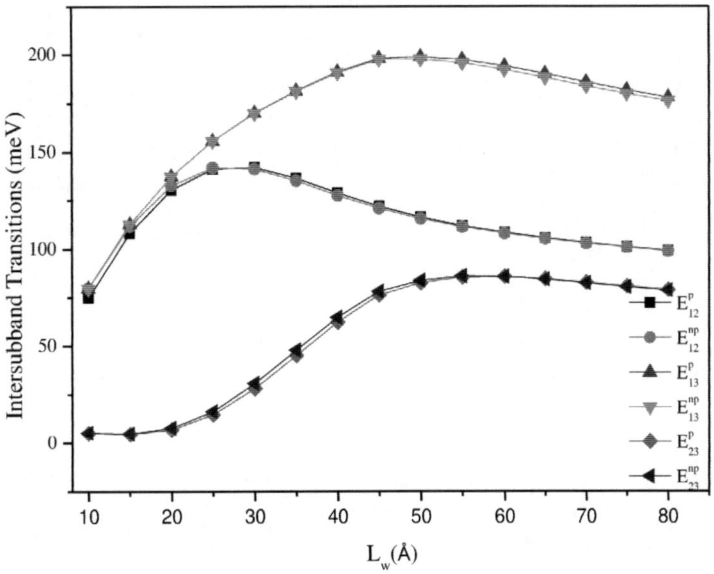

(b)

Figure 2. : (a) : Evolution confined levels depending the width of the well to L_b = 8nm. **(b)** : The transitions energies depending of the well width to L_b = 8nm.

For the nonparabolicity, its effect is negligible for the fundamental level. The E_{1p} and E_1^{np} are almost coinciding. But each time the level is excited the nonparabolicity effect becomes important; it lowers the confined level energy for wells immense, about ≈2.8eV for E_2 and ≈8eV for E_3. Thus, for fine wells the nonparabolicity effect is generally low.

From the figure 2. (b), it can be seen that each transition has two regions of variation, one where a transition increases with L_w, and the other where it decreases. In the 1st region, the upper level exists above Vs moves slowly, unlike the level below V_s decreases sharply, which resulting an increase in transition energy. In the second region, both levels have lower positions than V_s, the changes in transition energy has a usual behavior corresponding to a decrease, but this decreasing is lower than first. Between L_w = 10-20 Å, The transition E_{23} as a function of the well width remains practically constant. Indeed, the two Levels are located above V_s, where the displacement of the levels is low, with a variation almost similar.E_{12} calculations cover the range of the midinfrared wavelengths in the far infrared (λ_{12} = 9—16 µm).

Thus, those transitions have a maximum. The transition E_{13} has the highest, E_{13} = 194.5 meV for L_w = 50 Å corresponds to λ_{13}= 6.37 µm. concerning the field térahertzique the transition E_{23} allows us this emission/detection for the geometries with will width less than L_w= 35 Å, υ_{23} relaxes from 1.1 THz to 10 THz.

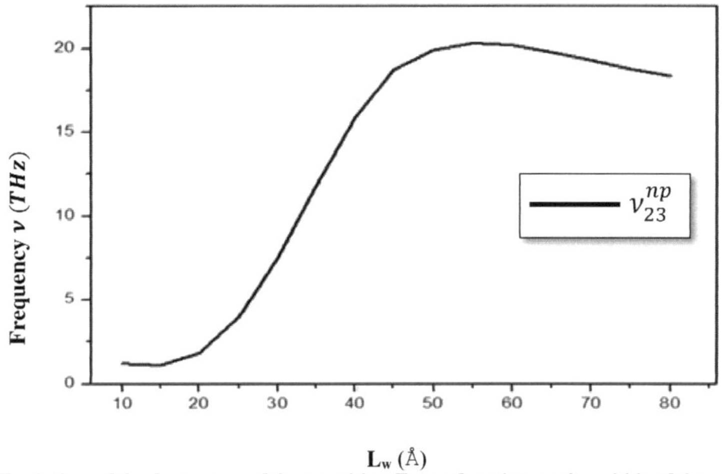

Figure 3. : Evolution of the frequency of the transition E_{23} as function as the width of the well.

The nonparabolicity reduces the transitional energies because of its effect on the confined levels positions, thus, for large wells, it drop the transition energy E_{13} by about ≈6 eV and the transition E_{12} by about ≈3 eV. However, for fine wells the effect of nonparabolicité is generally low, the transition energy E_{13} down by about ≈0.5 eV. Ours results are in good agreement with [19, 20, 21].

4. Transmission

4.1 Transmission in ZnO/Mg$_x$Zn$_{1-x}$O quantum well

We consider a ZnO / MgZnO heterostructure to study the tunnel effect through this structure as shown in the figure below.

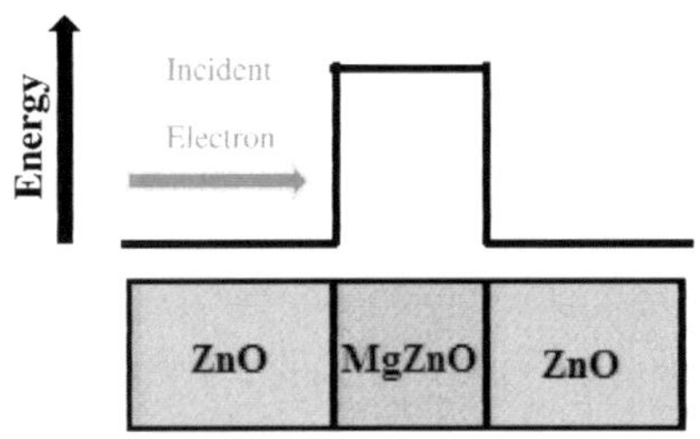

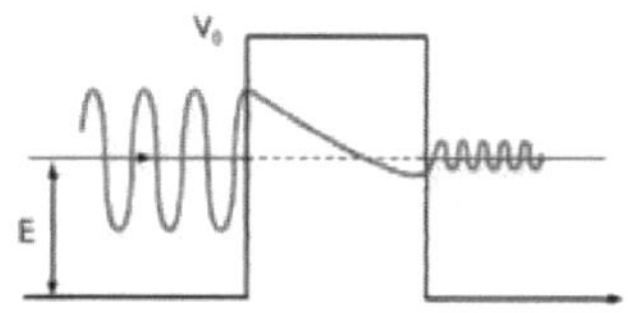

The transmission coefficient of an electron through the potential barrier is written as [19]:

$$T(E) = \frac{1}{1 + \dfrac{V_0^2}{4E(V_0 - E)} sh^2\left(\dfrac{L_b}{\hbar}\sqrt{2m(V_0 - E)}\right)} \tag{8}$$

With E the electron energy, V_0 the barrier height, L_b its width and m the electron mass.

Different variables are involved in the calculation of transmission coefficient: the barrier thickness, the electron energy or the height of the barrier. We have shown in the following figures(Figure 4. And Figure 5.), the influence of two parameters on the transmission coefficient: the thickness and height of the potential barrier.

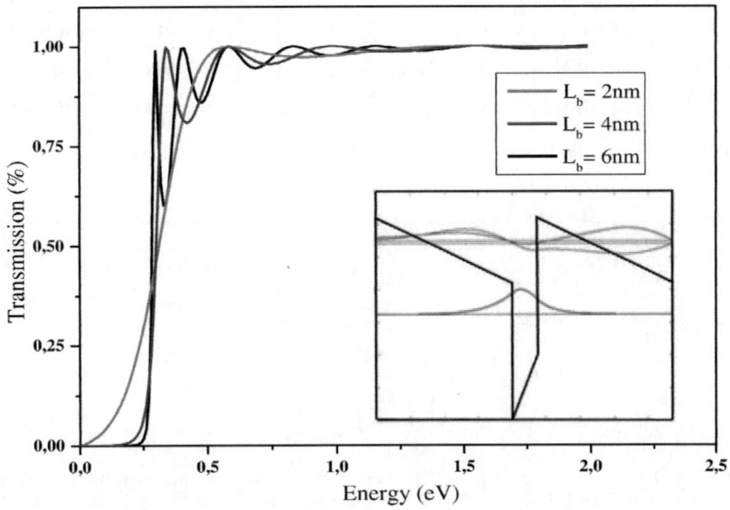

Figure 4. : The coefficient of transmission of a simple layer of MgZnO with 20% of Mg according to thickness of the layer with V_b=0.26 eV.

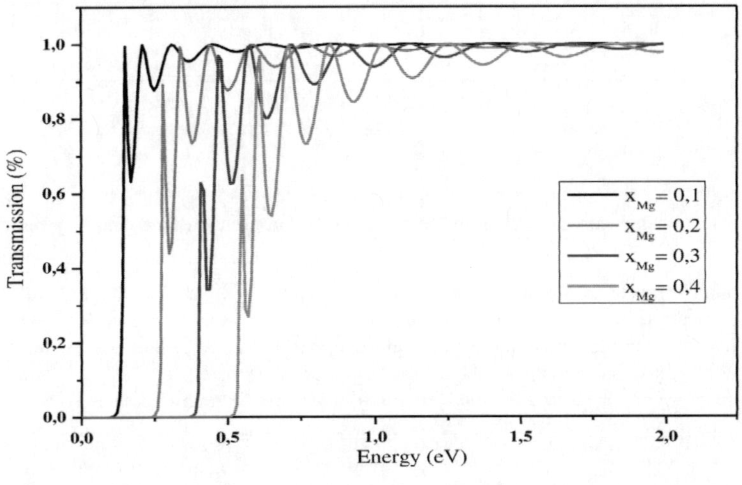

Figure 5. : The coefficient of transmission of a simple layer of MgZnO, thickness L_b=8 Nm According to X fraction of Mg.

Generally, in figure 4. the wider the barrier, the transmission of an electron with a given energy will be lower. When the thickness of the barrier is 5nm, very few electrons can pass through it (coefficient close to 0 when the energy is less than 0.26 eV). When the thickness is less than 5 nm, we see that the transmission coefficient is no longer zero for energies lower than the barrier height. The probability of tunneling increases. Beyond the barrier energy (0.2 eV), we observe on the thickest barrier, oscillations related to a phenomenon of quantum reflection.

Figure 5. Indicates that there is a decrease in the transmission coefficient when the barrier height increases (the fraction of Mg increases). In this case, it will be necessary to provide more energy to the electron so that it can cross the barrier. Beyond the barrier energy (V_b), we also notice variations related to a phenomenon of quantum reflection.

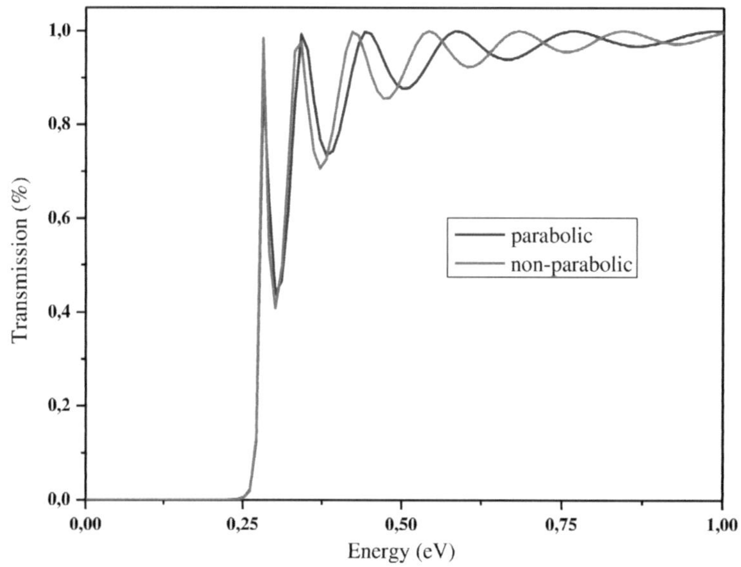

Figure 6. : The transmission coefficient in the parabolic and non-parabolic cases, L_b = 8nm And V_b = 0.526 eV.

The both samples present a good optical transmission around 90%, in most cases; thin multilayers have shown different physical properties. The variation of the shape of the individual MgZnO layers has a significant influence on the optical properties and the crystal structure of the MgZnO alloys. It can be seen that the transmission in the case of the parabolic bands are made in a fast way than in the non-parabolic bands. A good agreement is obtained with the existing literature values.

5. Conclusion

We have studied the conduction band profile and carried out calculations of the transitions energies in a quantum well ZnO/MgZnO with 20% of magnesium and for geometries 8/1--8/8 nm. Three levels of energy could be confined within a certain widths leading to three permissible transitions because the selection rules are changed (Asymmetric wells). We considered how the material parameters such as non-parabolicity affect those energies and wavelengths corresponding. The transition E_{12} cover the range of the mid-infrared wavelengths to the far infrared (λ_{12}=9—16 μm). Our results show that the transition energies E_{23} of the quantum well structures ZnO/MgZnO, which have very close mesh parameters, in advantageous to have an unconstrained structure, gave terahertzic frequencies ν_{23}= 1.1---10 THz used in the detection of explosives and safety in Airports. This serves to encourage the design and epitaxial growth of structures Photons. For the transmission coefficient, it depends strongly on the potential of the barrier (fraction of the magnesium) and its width of the well. It decreases when the height of the barrier increases and the barrier is fine. We also have enregistred the variations related to a quantum reflection phenomenon.

Reference

[1] Ü Özgür, Ya I Alivov, C Liu, A Teke, M A Reshchikov et al. *J. Appl. Phys.* **98**, 041301 (2005).

[2] S Nakamura, *Science* **281** (1998) *956*

[3] B Monemar, GPozina, Prog. *Quantum Electron.* **24** (2000) 239.

[4] Ü Özgür, Ya I Alivov, C Liu, A Teke, M A Reshchikov, S Dog˘an, V Avrutin, S J Cho, H Morkoç, *J. Appl. Phys.* **98** (2005) 041301.

[5] J Q Wu, *J Appl. Phys.* **106** (2009) 011101.

[6] F Benharrats « *Etude des nanostructures laser à base de ZnO/MgZnO* » Université d'Oran, (2007).

[7] S Singh, G R Dillip, S Vyas, M R Hasan, I K Park, P Chakrabarti, S H Park, Microsyst. Technol. **23** (2017) 39-46.

[8] C Morhain, X Tang, M Teisseire-Doninelli, B Lo, M Laügt, J M Chauveau, B Vinter, O Tottereau, P Vennégués, C Deparis, G Neu, *Superlatt. Microstruct.* **38** (2005) 455.

[9] M Yano, K Hashimoto, K Fujimoto, K Koike, S Sasa, M Inoue, Y Uetsuji, T. Ohnishi, K Inaba, *J Cryst. Growth* 301–302 (2007) **353**

[10] M W Allen, P Miller, R J Reeves, S M Durbin, *Appl. Phys. Lett.* **90** (2007) 062104.

[11] J A Davis, L V Dao, X Wen, C Ticknor, P Hannaford, V A Coleman, H H Tan, C Jagadish, K Koike, S Sasa, M Inoue, M Yano, *Nanotechnology* **19** (2008) 055205.

[12] Bhattacharya, P., Das, R.R., Katiyar, R.S., 2003. *Applied Physics Letters* **83**, 2010–2012

[13] M. Belmoubarik, K. Ohtani, H. Ohno, *Appl. Phys. Lett.* **92** (2008) 191906.

[14] K. Ohtani, M. Belmoubarik, H. Ohno, *J. Cryst. Growth* **311** (2009) 2176

[15] G Bastard « *Wave Mechanics Applied to Semiconductor Heterostructures* », les éditions de Physique, Paris (1988).

[16] N Grandjean, B Damilano, S Dalmasso, M Leroux, M Laügt, J Massies, *J Appl. Phys.* **86**, 3714 (1999).

[17] T Bretagnon, P Lefebvre, T Guillet, T Taliercio, B Gil, and C Morhain. Barrier composition dependence of the internal electric field in ZnO/ZnMgO quantum wells. *Appl. Phys. Lett.* **90,** 201912 (2007).

[18] H Mathieu and Hervet Fanet « *Physique des semiconducteurs et des composants électroniques* » 6ème édition, DUNOD (2009).

[19] A Rajira, H Akabli, A Almaggousi and A Abounadi, *Superlattices and Microstructures* **84** (2015) 192-197

[20] H Akabli, A Almaggousi, A Abounadi, A Rajira, K Berland and T G Andersson, *Superlattices and Microstructures* **52** (2012) 70-77

[21] H Akabli, A Almaggousi, A Abounadi, B Nacer and K Zekentes, *Phys. Chem. News* **49** (2009) 99-103

Growth and characterization of hybrid (HoGO/P3HT) graphene-based nanostructures for photovoltaic (PV) applications.

F V Molefe[1, 2], M Hamzah[1], M Khenfouch[1, 2], M S Dhlamini[1, 2] and B M Mothudi[1, 2]

[1]Department of Physics, UNISA, Private Bag X90, Florida, 1710, South Africa.
[2]Africa Graphene Centre (AGC), Department of Physics, UNISA, Private Bag X90, Florida, 1710, South Africa.

Abstract. Herein, we present a comparative study between nanostructures of poly(3-hexylthiophene) (P3HT), Holmium-Graphene Oxide (HoGO) nanocomposite and hybrid HoGO/P3HT thin-film nanostructures in terms of structural, morphological and spectroscopic properties. Specifically, the graphene based GO nanostructure was functionalized with rear earth ion Ho(III) to improve its mobility. Furthermore, semiconducting P3HT nanostructure was successfully grown with HoGO nanocomposite creating hybrid HoGO/P3HT nanostructure for energy materials. The nanostructures were characterized by X-ray diffraction (XRD), Scanning Electron Microscopy (SEM), (FTIR) and UV/VIS/NIR spectroscopy. The interaction between HoGO and HoGO/P3HT nanostructures is evidenced through substantial variations in nanoparticle morphologies. FTIR results provided the evidence of the presence of different types of carbon functionalities in the nanostructures. From the absorption spectra, growth of hybrid HoGO/P3HT nanostructure broadened the absorbance with a slight decrease in %. These nanostructures open a promising direction on growth of hybrids for photovoltaic applications because of their interesting optical properties.

Keywords: Hybrid, thin-film, nanocomposite, nanostructures, photovoltaics.

1. Introduction

For many years, intense scientific research has been performed on renewable energy materials that would replace conventional fossil fuels. The search for renewable and environmentally friendly energy sources is continuously increasing. As a result, many types of photovoltaic cells such as silicon, organic dyes, conducting polymers and graphene etc. as well as their combinations have been developed and tested [1]. Hybrid organic-inorganic thin-film nanostructures especially organic semiconductors comes in as materials of interest for tackling energy challenges we facing globally. In the class of hybrid nanostructures one of materials of interest is organic conducting polymers because they possess various chemical structures.

Semiconducting polymers gained more interest because it is easy and cheap to process them at large scale from solution, provided they have impressive optoelectronic properties. Polythiophenes, in particular poly(3-hexylthiophene) (P3HT), are the most versatile representatives of polymer semiconductors investigated for various applications such as solar cells, supercapacitors and field-effect transistors [2,3]. It has been well established that tuning the surface morphology of this class of polymers with functional layer of nanomaterials has a significant impact on the performance of

fabricated devices. However, it remains a challenge to determine charge transportation due to the complex semi-crystalline microstructure which consists of crystalline and amorphous domains. One way to improve charge transport is to make use of new materials with large surface area as active components. In this context, graphene is the most suitable candidate, the large surface area of which could be used to accept transported charges. In the previous research outputs, conducting polymers and graphene have been synthesized separately and mixed together in the solution to form composite structure [4, 5], but due to re-stacking property of graphene oxide layers the performance of photovoltaic devices was poor. Then we considered developing thin-film nanostructures where we've got GO mixed with Holmium (Ho) to create HoGO nanocomposite.
Ho is a rare earth ion known to exhibit several interesting properties such as large magnetic moments and luminescence. The latter are shown in photostable upconversion nanoparticles, where energy transfer between two elements (e.g., ytterbium and erbium) give rise to multicolor luminescence [6]. Ho is within class of elements that possess unique optical, catalytic and magnetic properties due to the unique configuration of their 4f electrons. Ho (III) possess one of the highest magnetic moments of all elements, but of interest to our study is its spectroscopic characteristics. Introducing these atoms into a different matrix can lead to a material that shows multiple interesting effects. This means, its presence can greatly prevent the charge recombination of photo-induced electrons and holes in semiconductors and enhance photocatalytic activity and photosensitivity.

Our work paves the way to growth of hybrid HoGO/P3HT nanostructure leading to design of devices as potentially efficient for PV applications. Structural, morphological and optical properties were probed to clarify functionalization and interaction between the layers of nanostructures. The interaction between hybrid HoGO/P3HT nanostructures is important in the development and design of new solar cells devices.

2. Experimental

2.1. Materials Preparation

Preparation of P3HT is similar to the one we stated in our previous report [7]. For the preparation of GO, we have adopted the modified Hummers method presented by Khenfouch *et. al.* [8]. Upon preparation of Holmium (Ho) solution in 50 ml of distilled water we added 0.1g NaOH, the solution was stirred for 10 min until NaOH was completely dissolved. After that, a solution of 0.5g of Holmium (III) Nitrate $Ho(NO_3)_3$ in 40 ml of distilled water was prepared and stirred using a magnetic stirrer for 30min. Later, a volume of 10 ml from the NaOH solution, was mixed with 40 ml from the $Ho(NO_3)_3$ solution followed by heat treatment in an oven at 90 °C for 24 hours. The inclusion complex for making HoGO nanocomposite was prepared by mixing the solutions in the molar ratio 1:1 Ho:GO and stirred for 30 min by ultrasonication. For all samples under investigation, 1.2 x 1.2 cm^2 glass substrates were cleaned successively in deionized water, ethanol and acetone for 15 min prior coating. The P3HT solution was preheated to 50 °C for 15 min prior coating, in order to prevent polymer pre-aggregation and gelation in solution [9]. Hybrid thin-film nanostructures were prepared by drop casting solutions on glass slide substrates. Right after deposition, thin-film nanostructures were dried in an oven at 50 °C for period of 2hrs to remove residues.

2.2. Materials Characterization

The phase and purity of thin-film nanostructures were examined by X-ray diffraction in Reflection (Bragg-Brentano geometry) by a Rigaku Smart Lab diffractometer with CuKα (1.5418Å) radiation. Morphological study was done using the Field Emission Scanning electron microscope (FE-SEM) (Shimadzu model ZU SSX-550 Superscan). FTIR PerkinElmer spectrometer in the scan range 400 to 4000 cm-1 was used to probe chemical bonding. The optical absorbance of the thin-films in the UV to visible region (250-800 nm) were performed with (PerkinElmer Lambda 1050 UV/VIS/NIR Lambda) spectrometer equipped with an integrating sphere.

3. Results and discussion

3.1. XRD analysis

Figure 1 presents the XRD patterns of nanostructures, P3HT, HoGO and HoGO/P3HT. The XRD diffraction pattern of P3HT reveals an amorphous basis dominated by (100) peak at ~5.7 ° corresponding to its well-known diffractions from literature [7]. Additionally, there are other diffraction peaks represented with asterisk (*) that can be attributed to the glass substrate. The HoGO nanocomposite presented glass substrate peak at ~8.7 ° which is also observed in P3HT and at ~10.3 ° there is an emerging diffraction indicating the existence of GO with couple of small development of Ho phase. In the case of hybrid HoGO/P3HT nanostructure, the diffraction patterns become broader suggestive of decrease in crystallinity. Zhao et. al. [10] indicated that crystallization of P3HT is hampered by the crystalline entities when interacting with other nanostructures leading to less ordering of the semiconducting moieties.

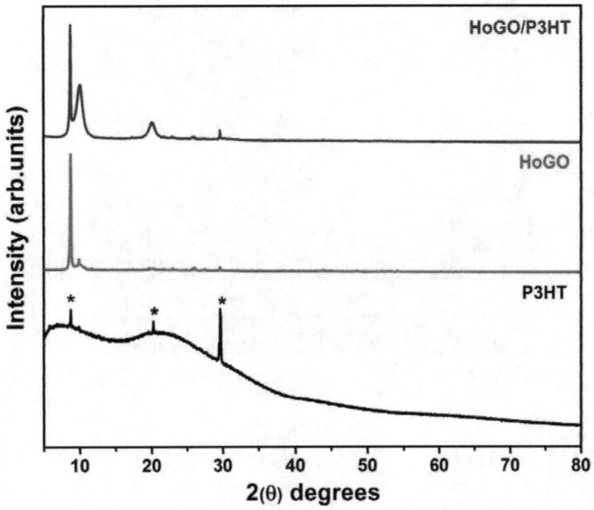

Figure. 1 XRD spectra of P3HT, HoGO and HoGO/P3HT nanostructures.

3.2. SEM analysis

In **Figure 2(a)** we have SEM image of pristine P3HT nanostructures showing clusters of small spherical particles. It can be seen that the distribution of particles with respect to the substrate is not uniform throughout the film. This is caused by the substrate being highly sensitive to the polymer film thickness [11]. The clustering symbolizes that π-stacked aggregates are oriented and distributed in an amorphous matrix. It is well explained in literature that surface morphology of regioregular P3HT is highly affected by its parameters such as molecular weight, regioregularity and polydispersity [12, 13]. **Figure 2(b)** presents morphology for HoGO nanocomposite wherein we observed some wrinkles from GO sheets and rod-like structures representing the inclusion of Ho^{3+} ions. The SEM image in **Figure 2(c)** for hybrid HoGO/P3HT nanostructure indicated a relatively smooth surface and small islands. This shows that diffusion was undertaken between P3HT and HoGO leading to internal and surface interaction. Our results agree with statement by Tian *et. al.* [14] that organic compounds including both polymers and small molecules have more compatibility and have been widely used as components to tune the morphology.

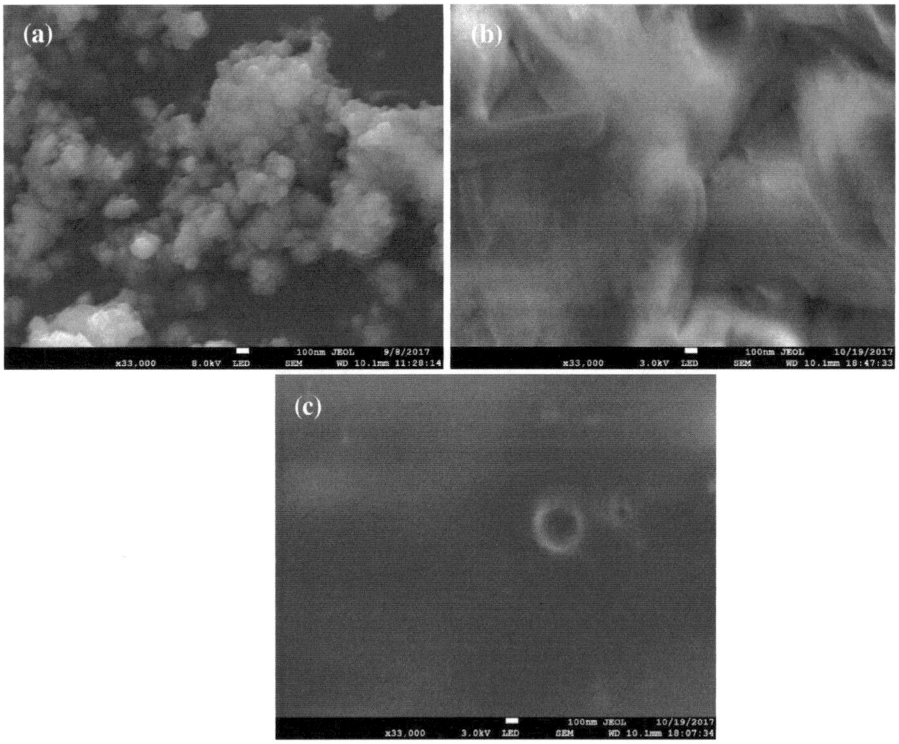

Figure. 2 SEM images of (a) P3HT, (b) HoGO and (c) HoGO/P3HT nanostructures.

3.3. FTIR analysis

Figure 3 shows the FTIR spectra helpful for determining various vibrational frequencies of P3HT, HoGO and HoGO/P3HT nanostructures in the range 4500-500 cm^{-1}. In all nanostructures we observed carbon peak at ~2364 cm^{-1} coming from environmental CO_2 because it is hygroscopic. The spectrum for regioregular P3HT exhibits principal vibration bands at ~671, ~819, ~1035 and ~1307 cm^{-1} indicating aromatic out of plain =C-H vibration band, which is the characteristic of 2.5-disubstituted-3hexylthiophene for rr-P3HT chain conformation [15]. Other C-H vibration band is seen at ~1460 cm^{-1} which is associated with symmetric stretching vibration as well as aliphatic C-H stretching bands at ~2856 and ~2924. Furthermore, there's a weak band associated with C=O stretching vibration at ~1750 cm^{-1} verifying the presence of ester group and it appears to be the most pronounced for hybrid HoGO/P3HT nanostructures. HoGO nanocomposite only revealed weak bands located at ~2000 and ~2171 cm^{-1} corresponding to the oxygen containing groups on the surface. Main changes were observed at lower side of FTIR spectra for hybrid HoGO/P3HT nanostructure where four strong characteristic adsorption peaks appear at around ~1241, ~1362, ~1436 and ~1750 cm^{-1}, respectively. These peaks correspond to C-H vibration band which indicates partial decomposition of the thiophene [16]. We observed ~3267 cm^{-1} peak indicating water absorbed on the surface of P3HT and this vibration band is not applicable in HoGO and HoGO/P3HT. The significant reduction in vibrational frequencies observed in hybrid HoGO/P3HT nanostructure may be due to oxygenation at the interface.

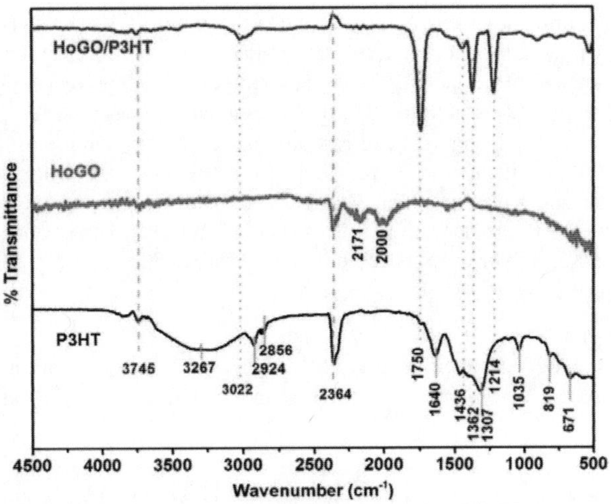

Figure. 3 FTIR spectra of P3HT, HoGO and HoGO/P3HT nanostructures.

3.4. UV/VIS/NIR analysis

Figure 4 represents absorption spectroscopy measurements, which is helpful for determination of degree of intra and interchain order within thin-film nanostructures containing P3HT. Firstly, pristine P3HT have slightly better absorption in the region 300-600 nm. The UV-vis spectrum of HoGO nanocomposite has a lower absorption compared with P3HT with well-defined shoulders except the 574 nm. The growth of hybrid HoGO/P3HT nanostructure enhances the absorption and there is better enhancement in the 600-800 nm region. Furthermore, the absorption spectrum of HoGO and HoGO/P3HT took the same shape as evidence of interaction with the rebuilt 574 nm absorption band of pristine P3HT on hybrid HoGO/P3HT nanostructure. This is because when the thin-film is completely dried, the arrangement of organic materials does not change. Some researchers indicated that absorption peaks tends to shift to the blue or red region when the interchain interaction between organic materials is changed [15, 17]. However, we observed no wavelength shift in the absorption spectra indicating that the interchain interaction is not responsible for improved absorption. We assign the absorption enhancement to the increase in internal reflection caused by modified arrangement of organic materials in thin-films. The improved absorption in 600 - 800 nm for hybrid HoGO/P3HT nanostructures may be due to highly delocalized electron rich and few charges formed along the π-conjugated thiophene backbone upon growth of nanostructure [18].

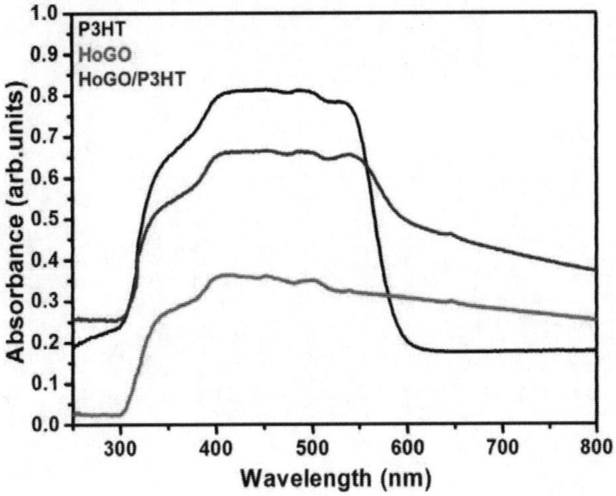

Figure. 4 UV-Vis spectra of P3HT, HoGO and HoGO/P3HT nanostructures.

4. Conclusion

It was feasible to grow thin-film nanostructures of hybrid HoGO/P3HT. Through the XRD results we can declare that the synthesized nanostructures revealed phases of P3HT, HoGO and HoGO/P3HT. The surface morphology of hybrid HoGO/P3HT nanostructure became very smooth with some tiny islands and there was no indication of spherical, rodlike and nanosheets observed in P3HT and HoGO before growing film onto another. The evidence of chemical interaction between thin-film nanostructures is seen from changes observed in FTIR stretching/vibrational modes. From optical study the nanostructures present a broader absorption for the hybrid over a larger spectral range towards IR region which will enhance the performance of devices. These finding are useful in the fabrication of graphene based hybrid nanostructures for PV applications.

Acknowledgements

We would like to acknowledge financial support from National Research Foundation (NRF) of South Africa and UNISA College of Science Engineering and technology (CSET) and Africa Graphene Center (AGC).

References

[1] R. Liu, *Materials*, **7** (2014) 2747.

[2] R. Zhang, B. Li, M.C. Iovu, M. Jeffries-EL, G. Sauve, J. Cooper, S. Jia, S. Tristram-Nagle, D.M. Smilgies, D.N. Lambeth, R.D. McCullough and T. Kowalewski, *J. Am. Chem. Soc.* **128** (2006) 3480.

[3] G. Li, V. Shrotriya, J. Huang, Y. Yao, T. Moriarty, K. Emery and Y. Yang *Nat. Mater.* **4** (2005) 864.

[4] Y. Wang, S. Yang, H. Wang, L. Zhang, H. Cheng, B. He, W. Li and B. Zou, *Curr. Appl. Phys.* **17** (2017) 343.

[5] Z. Yang, X. Shi, J. Yuan, H. Pu and Y. Liu, *Appl. Surf. Sci.* **257** (2010) 138.

[6] F. Vetrone, R. Naccache, V. Mahalingam, C.G. Morgan and J.A. Capobianco, *Adv. Funct. Mater.* **19** (2009) 2924.

[7] F.V. Molefe, M. Khenfouch, M.S. Dhlamini and B.M. Mothudi, *Advanced Mater. Lett.* **8** (2017) 246.

[8] M. Khenfouch, U. Buttner, M. Baitoul and M. Maaza, *Graphene* **3** (2014) 7.

[9] P. Ehrenreich, S.T. Birkhold, E. Zimmermann, H. Hu, K.-D. Kim, J. Weickert, T. Pfadler and L. Schmidt-Mende, *Scientific Reports* **6** (2016) 32434.

[10] K. Zhao, Zi. Ding, L. Xue and Y. Han, *Macromol. Rapid Commun.* **31** (2010) 532.

[11] S. Joshi, S. Grigorian, U. Pietsch, P. Pingel, A. Zen, D. Neher and U. Scherf, Macromolecules **41** (2008) 6800.

[12] Z. Wu, A. Petzold, T. Henze, T. Thurn-Albrecht, R.H. Lohwasser, M. Sommer and M. Thelakkat, *Macromolecules* **43** (2010) 4646.

[13] A. Zen, J. Pflaum, S. Hirschmann, W. Zhuang, F. Jaiser, U. Asawapirom, J.P. Rabe, U. Scherf and D. Neher, *Adv. Funct. Mater.* **14** (2004) 757.

[14] H. Tian, G. Boschloo and A. Hagfeldt, *Green Chemistry and Sustainable Technology*, Springer.

[15] R.K. Singh, J. Kumar, R. Singh, R. Kant, S. Chand and V. Kumar, *Materials Chemistry and Physics* **104** (2007) 390.

[16] P. Kovacikand and O. College, "Vacuum Deposition of organic molecules." PhD thesis University of Oxford.

[17] C.-X. Sheng and Z.V. Vardeny, *Fundamentals Materials devices and Fabrication*, ISBN:978-3-662-45508-1 (2015) 3.

[18] R.K. Singh, J. Kumar, R. Singh, R. Kant, R.C. Rastogi, S. Chand and V. Kumar, *New J. Phys.* **8** (2006) 112.

pH and concentration effect on the optical absorption properties of Sn(V) tetrakis (4-pirydyl) porphyrin functionalized graphene oxide.

O. Bajjou[1,3] , A. Bakour[1], M Khenfouch [2,3] , M Baitoul[1] , B Mothudi[2], M Maaza[4,5] and E. Faulques[6]

[1]University Sidi Mohamed Ben Abdellah, Faculty of Sciences Dhar el Mahraz, Laboratory of Solid state Physics, Group of Polymers and nanomaterials, PO Box 1796 Atlas Fez 30 000, Morocco.
[2]Physics department, College of Science, Engineering and technology Science Campus, Corner of Christiaan de Wet Road & Pioneer Avenue, Florida, 1709, Johannesburg, South Africa
[3]Africa Graphene Center, Physics Department, CSET, Florida Campus, University of South Africa, Johannesburg, 1710, South Africa
[4]iThemba LABS-National Research Foundation of South Africa, Old Faure Road, PO Box 722, Somerset West 7129, Western Cape Province, South Africa.
[5]UNISA Africa Chair in Nanosciences-Nanotechnology, College of Graduate Studies, University of South Africa, Muckleneuk ridge, PO Box 392, Pretoria, South Africa.
[6]Institut des Matériaux Jean Rouxel, Nantes, 2 rue de la Houssinière, BP32229, 44322 Nantes, Cedex3, France.

Email: bajjou.omar@gmail.com

Abstract. The graphene oxide functionalized Sn(V) tetrakis (4-pirydyl) porphyrin [$SnTPyP^{2+}$] nanostructure, was prepared by simple method with multiple GO concentrations and different acidity levels. The vibrational properties were investigated by means of Raman scattering and FT- Infrared spectroscopies, and showed that electronic interactions take place at the interface. At a structural level the X-Ray Diffraction [XRD] analysis revealed the effect of the $SnTPyP^{2+}$ on the GO sheets interspacing. The optical absorption was studied for many pH values and GO concentrations, and led to confirming the existence of strong interaction between both components of the composite GO-$SnTPyP^{+2}$. The resulting GO-$SnTPyP^{2+}$ was proved to be pH-sensitive in the absorption of photons, and hence lead to choose the best composite for efficient electron/energy transfer between GO and the porphyrin, and affords improved optical nonlinearities compared to those of the GO and the pristine porphyrin.

1. Introduction

Graphene, a single atomic layer of sp^2-bonded carbon atoms arranged in a honeycomb lattice [1], exhibits remarkable electronic, thermal and mechanical properties [2]. A variety of applications have been demonstrated for graphene materials, such as structural nanocomposites [3-5], catalyst supports, electronic devices, and energy storage in batteries and supercapacitors [6]. One of the most promising methods for large scale production of graphene is by chemical oxidation exfoliation of graphite leading to graphene oxide (GO)[7], which can subsequently be reduced to graphene. GO has generated significant interest due to its ability to form stable colloids in water and other solvents as well as in different matrices due to the presence of carboxylic and hydroxyl groups[8, 9]. Furthermore, graphene's strong interactions with photons and electrons, and chemical functionalization ability could add more functions to photoactive composites[10, 11].

Porphyrins have features that allow molecular and π-conjugated surfaces to gather in the form of the tetrapyrrole macrocycles. Porphyrins are very attractive molecules in many aspects. The first reason is related to their unique spectroscopic properties both in absorption and emission. Porphyrins and related tetrapyrroles have been extensively studied because of their importance in biological processes and they are often used in the development of artificial light-harvesting, energy conversion[12, 13]. They are also interesting from a fundamental point of view, as systems to study the nature and dynamics of excited states in molecular assemblies of reduced dimensionality, as well as models for understanding the photophysics of natural light-harvesting complexes. The extraordinary optical properties of these molecular aggregates are to a large extent governed by exciton delocalization, transport, and relaxation phenomena[14]. To develop these properties, composites based on those structures show their potentials for optical and optoelectronic applications by the association of porphyrin properties to carbon nanotubes, polymers and other molecular materials [15-17].

Another interesting aspect is the functionalization of few layers of GO with Sn(IV) tetrakis(4-pyridyl) porphyrin dichloride ($SnTPyP^{2+}$), by van der Waals and/or electrostatic forces. Previous reports have demonstrated that GO has the excellent and tunable hydrophilicity. However, few reports have focused in detail on the influences of particle size and pH value on the hydrophilicity of GO. Herein, we investigate the impacts of pH value on the GO- $SnTPyP^{2+}$ composites.

In this paper, we studied the vibrational and structural variations that affect the composite, and also the effect of pH on the absorption properties of graphene oxide (GO) when incorporated into $SnTPyP^{2+}$. X-ray diffraction, FTIR, Raman spectroscopy and optical absorption were used in order to investigate the structural and electronic properties of this system for optoelectronic applications.

2. Materials setup

using Hummer's method we succeeded to obtain Graphene oxide from graphite [18]. The GO solution was prepared using ($KMnO_4/H_2SO_4$), and then the precipitate was washed many times with de ionized water. After obtaining a grey solution of GO, we prepared many pH concentrations of the solution, starting from an acidic level of 2 rising up to 10, and then added to already prepared $SnTPyP^{2+}$ solutions. For optical characterization we prepared also different concentrations of the GO sample. The final GO solutions were diluted and sonicated for a major exfoliation of the oxidized sheets. The solutions were then submitted to a dropcasting and drying process to obtain homogeneous films.

3. Characterization

Raman spectra were obtained using a Bruker spectrometer SENTERRA Raman Stokes and spectral range of 90-3200 cm^{-1} with an excitation wavelength $\lambda = 514$ nm and for our samples the laser power is adjusted to a low power 10 mW. FTIR spectra were recorded at room temperature in the transmittance mode using a Vertex 70 spectrophotometer, with a resolution of 4 cm^{-1} over a wavenumber range of 400-4000 $cm\text{-}^{1}$. UV–Vis spectra were measured on a PerkinElmer LAMBDA950/1050 UV/VIS/NIR; spectrophotometers are dual-beam; dual-monochromator instruments equipped with an optional 150 mm integrating sphere accessory.

Time-resolved photoluminescence (TR-PL) experiments were acquired with a regenerative amplified femtosecond Ti: Sapphire laser system (Spectra Physics Hurricane X). This setup generates 100 fs pulses at 800 nm with a repetitive rate of 1 kHz and a power of 1 W. The laser line is frequency-doubled (tripled) with thin BBO crystals to obtain an excitation line λexc= 400 nm (3.1 eV). The pump energy pulse is controlled to ensure that the excitation density in the sample did not exceed 10 17cm3, to avoid bimolecular annihilation process and sample photodegradation. The emission is temporally resolved with a high dynamic range Hamamatsu C7700 streak camera coupled to an imaging spectrograph with a temporal resolution to be estimated 20 ps and processed using the HPDTA Hamamatsu software.

4. Results & discussion

4.1 Raman measurement

Raman spectroscopy is a non-destructive technique that is widely used to obtain structural information of nanocarbons. In Figure.1 the Raman spectrum of GO presents a D-band at 1352 cm^{-1} associated to defects and disorders in carbon nanostructure. The G-band appearing at 1588 cm^{-1} arises from the E_{2g} phonon modes of graphite in the Γ-point and it is originating from the stretching of the C-C bond in graphitic materials, and is common to all sp^2 carbon systems [19, 20]. Meanwhile, the D peak which is also present due to the breathing mode of aromatic rings, which comes from the structural imperfections created by the attachment of oxygenated groups on the carbon basal plane. Therefore, the D-peak intensity is often used as a measure for the degree of disorder [21-23].

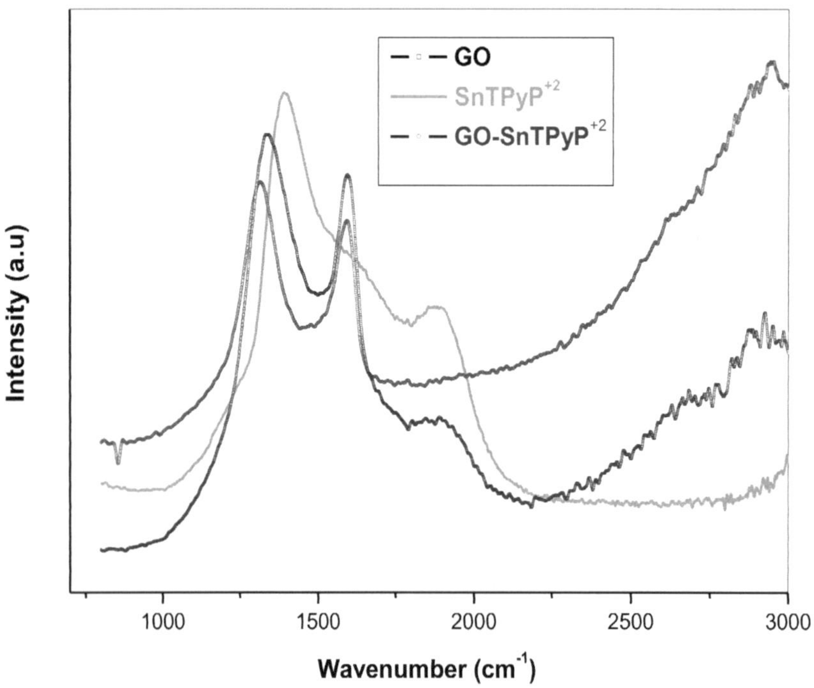

Figure 1. Raman spectra of GO, $SnTPyP^{2+}$ and GO– $SnTPyP^{2+}$

The functionalized GO with various porphyrins results in small changes in the D, G and 2D peaks. When the Raman spectra of GO– $SnTPyP^{2+}$ are compared with that of GO, subtle rather than drastic changes were observed. The G-band up shifted from 1588 to 1595 cm^{-1}, which might be caused by the increasing number of layers in their solid states. As porphyrins are covalently linked to GO, they can absorb on both sides of the graphene sheets. Moreover the 2D band intensity decreases when compared with that of GO. Generally, the integrated intensity ratio of the D and G-bands (ID/IG) indicates the oxidation degree and the size of sp2 ring clusters in a sp3/sp2 hybrid network of carbon atoms [24]. Moreover, the intensity ratio of D/G bands increased from 0.91 for GO to 0.96 for GO-SnTPyP2+. These changes indicate the presence of π-π interaction between the GO and $SnTPyP^{2+}$.

4.2 FT Infrared measurements

FT-IR spectroscopy is recognized as an important tool to study different types of functional groups. In Figure 2, the FT-IR spectra of GO, $SnTPyP^{2+}$ and GO- $SnTPyP^{2+}$ composite are exhibited. In the Go spectrum, we note firstly, the peak at 1408 cm^{-1} is assigned to the C–O vibrational mode. Secondly, the band centered at 1616 cm^{-1} is attributed to aromatic C=C bonds. Another peak at 1715 cm^{-1} corresponds to the vibrational mode of the ketone groups C=O. Finally, the peak at 3200 cm^{-1} denotes C–OH stretching [25, 26].

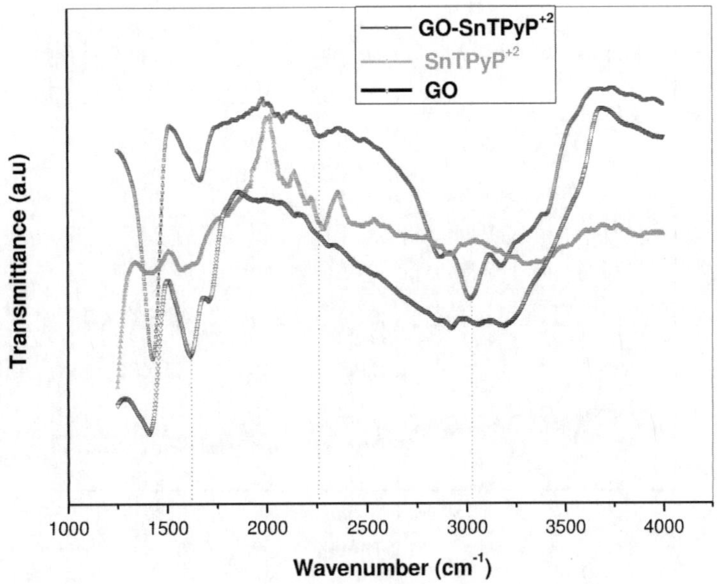

Figure.2: FTIR spectra of GO, $SnTPyP^{2+}$, and GO-$SnTPyP^{2+}$

For $SnTPyP^{2+}$ we note the presence of a broad intense band centered at 3400 cm^{-1} which is attributed to the N-H bond stretching, and the band situated at 2914 cm^{-1} is related to stretching vibration of aromatic C-H bonds. In addition, a band was observed at around 1629 cm^{-1} corresponding to the stretching bond in the phenyl nuclei. Moreover, the band at 1540 cm^{-1} is attributed to C_β-$C_{\beta'}$ stretching which may be due to symmetrical deformation of the pyrrole rings and the bending strain C_α- N- $C_{\alpha'}$. The band at 1415 cm^{-1} is attributed to the symmetric stretching C_α- N- $C_{\alpha'}$ and the C_β-$C_{\beta'}$ bond stretching in pyrrole nuclei, there is also a band which appears at 1318 cm^{-1} corresponding to the asymmetric HN-C_α-C_β stretching bond. The bands at 1261 cm^{-1} and 1298 cm^{-1} are assigned to the asymmetric C_α-N-$C_{\alpha'}$ stretching [27, 28]. In GO- $SnTPyP^{2+}$ we find that the intensities of these bands change significantly. Moreover, the vanishing of the band at 1408 cm^{-1} indicates a strong interaction between GO and $SnTPyP^{2+}$ which can be explained by covalent functionalization in GO-$SnTPyP^{2+}$[29].

4.3 X-ray diffraction

Figure.3 shows the XRD patterns of GO, $SnTPyP^{2+}$, and GO- $SnTPyP^{2+}$. GO spectrum exhibits an intense and strong peak at $2\theta = 11.10°$ which is attributed to the (001) lattice spacing of 0.78 nm due to the interlamellar water trapped between the hydrophilic graphene oxide layers [30-33]. For $SnTPyP^{2+}$ the XRD patterns shows a peak at $2\theta=8.8°$ attributed to the (001) diffraction plan and another peak at $2\theta =43.75°$ attributed to the (701)[34, 35]. The GO-$SnTPyP^{2+}$ composite pattern shows the characteristic peak (001) of the GO shifted to $2\theta=10.60°$ (d-spacing = 0.881 nm). In addition, the intensity of all diffraction peaks corresponding GO decreased after the interaction with the $SnTPyP^{2+}$, this behavior can be explained by the increase of the inter-planar distance (001) due to the effect of $SnTPyP^{2+}$ trapped by GO sheets.

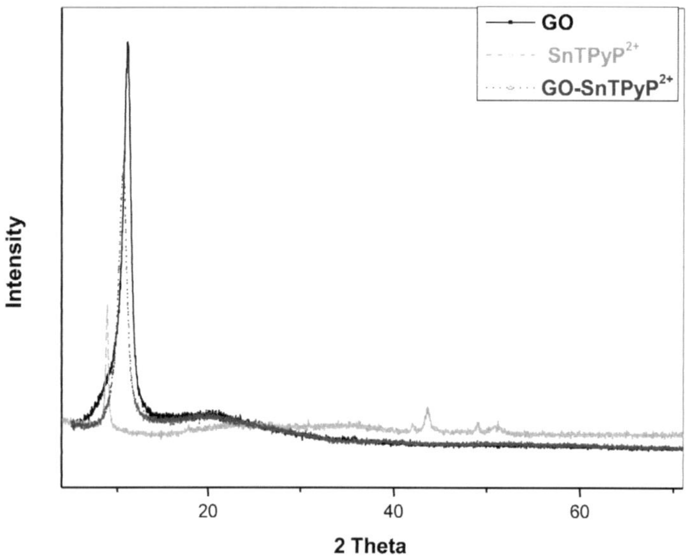

Figure 3. XRD of GO, $SnTPyP^{2+}$, and GO- $SnTPyP^{2+}$

4.4 UV-Visible Absorption

UV-Vis absorption spectra of the different samples are shown in Figure. 4. The absorption of GO, Figure 4(b), exhibit two peaks centered at 230 and 300 nm are observed for the GO dispersion, corresponding to the π-π^* transition of aromatic C=C bonds and the n-π^* transition of the C=O bond, respectively [8, 36]. we note a widening of the band characteristic of GO situated at 230 nm and broad band at 308 nm when going from concentration 0.1% to 1% and 2% and also we can observe an increasing in the intensity. This intensity changing and widening of the bands associated with the increase in concentration can be understood in terms of the increase in the size of the graphene layer.

In figure.4(a), we show the variations occurring at the absorption spectrum when changing the pH values of GO. The nature of the carboxylic, hydroxyl and epoxy group of GO also changes, causing a different degree of passivation of the GO surface. For pH<=4, the GO is fully protonated, and at pH > 4, the anchored–COOH groups become deprotonated such that the intensity decreases significantly

[37, 38]. The broad peak absorbance situated at 304 nm decreased with increasing pH explained the effect of solvent and the NaOH or HCl concentration[39].

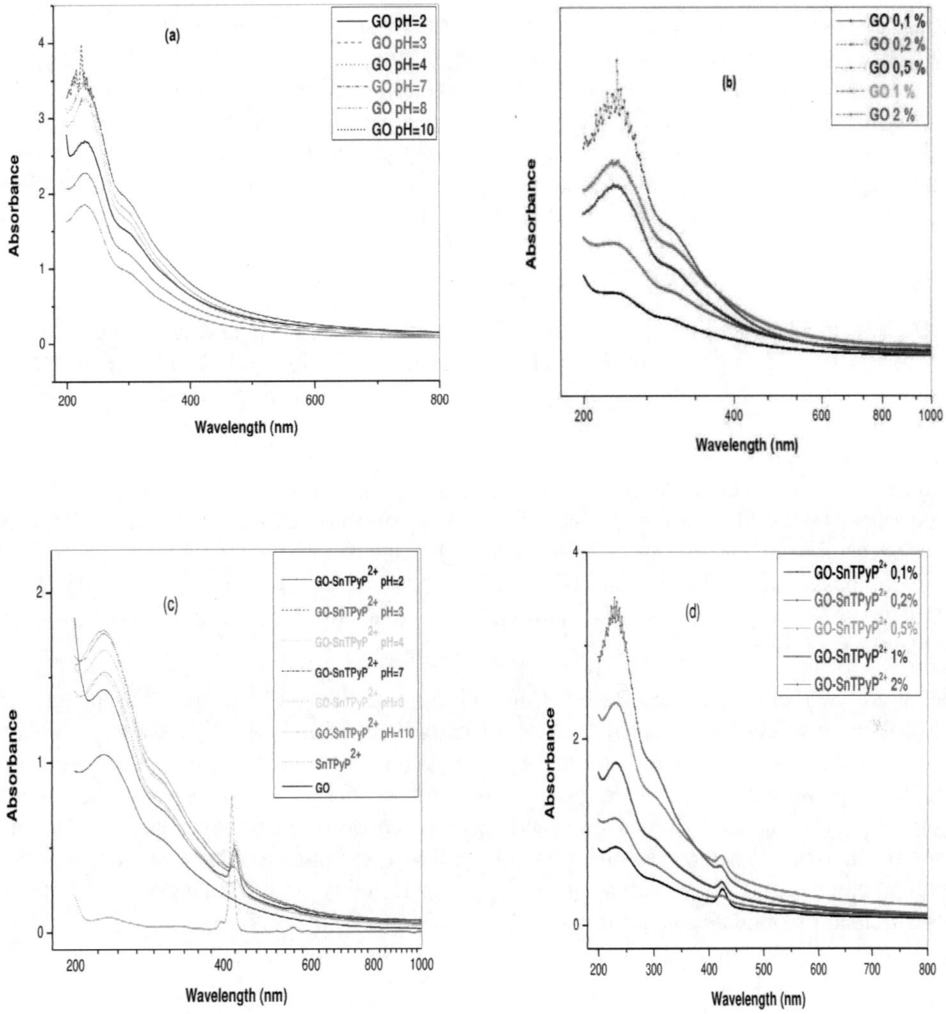

Figure 4. Optical absorption of GO, and GO- $SnTPyP^{2+}$; (a) and (c): pH effect; (b) and (d): concentration effect

The UV-Visible absorption spectrum of $SnTPyP^{2+}$, Figure 4(c), consists of two distinct regions. In the violet region, an extremely intense absorption known as the Soret or B band situated at 416 nm in the visible region, a number of absorptions known as the Q bands are present, two major absorption bands situated at 554 nm and 591 nm attributed Q(0.0) and Q(1.0) respectively[40]. The lower energy band, Q(0,0), is attributed to excitation to the first excited state. The higher energy band, Q(1,0), is a result of vibrational transitions interacting with the electronic transitions of the porphyrin, and has been referred to as a vibronic overtone of the Q(0,0) band. The Soret band is attributed to excitation to the second excited state, and in well resolved spectra a vibronic overtone, B(1,0). Both of these transitions are π–π^* transitions that involve electron redistribution within the conjugated π systems of the porphyrin rings[41].

The interactions between GO and $SnTPyP^{2+}$ in such a system are mainly electrostatic in origin, and energy is transferred non-radiative from the excited state of one molecule to the next by a long range dipole-dipole coupling mechanism. The energy of the exciton state in a porphyrin is dependent on whether the dipole interaction is attractive or repulsive, and an exciton state is only allowed if the dipoles are in-phase. Moreover, two mechanisms for energy transfer are generally anticipated to operate in the singlet states; Dexter and Förster. The former mechanism involves a double electron exchange between the donor in its singlet excited state and the acceptor in its ground state (LUMO (donor*) →LUMO (acceptor) and HOMO (donor*)←HOMO(acceptor)) and consequently strongly depends on the degree of orbital overlap between the donor and acceptor.

After $SnTPyP^{2+}$ was incorporated with the GO at different pH values, we note a widening of the band characteristic of GO when going from pH2 to pH10 increasingly moving spikes to low energies. This redshifted with the increase in pH can be understood in terms of the restored electronic conjugation pulling the highest occupied molecular orbital and lowest unoccupied molecular orbital. Moreover, GO-$SnTPyP^{2+}$ we observe a change in the intensity in the composite when passing from PH2 to PH10 can be explained the intramolecular donor-acceptor structure usually allows charge-transfer interaction or the highest occupied molecular orbital (HOMO)-lowest unoccupied molecular orbital (LUMO) transitions to occur. Moreover, in the visible region the band around to 380 nm of $SnTPyP^{2+}$ disappeared in composite GO-$SnTPyP^{2+}$ and the redshift B-band when going from pH2 to pH10 can be explained by decrease in oxygen functional groups and an increase in aromatic rings, causing electrons to be easily excited at a lower energy.

These results prove interaction between the electron and the hole is strong and the excitons tend to be localized almost exclusively on porphyrin. The efficiency of this energy transfer is inversely proportional to the sixth power of the distance between donor and acceptor. So, the energy transfer rate depends on the strengths of the electronic transitions for donor and acceptor molecules, and requires resonance between donor fluorescence and acceptor absorption. Energy transfer between GO and $SnTPyP^{2+}$ is known as heterogeneous transfer, unlike the homogeneous transfer that occurs between identical molecules. Direct evidence of energy transfer between different composite (GO and $SnTPyP^{2+}$) is provided by optical absorption.

4.5 Photoluminescence measurements

In this study, the excitation wavelength is 420 nm, in order to be closer to the resonance of the porphyrin molecules. We would then see if excitation resonant with transitions of the molecule allows the same photostability of deposit. Figure 5 shows the steady-state photoluminescence spectra of GO, $SnTPyP^{2+}$ and GO-$SnTPyP^{2+}$.

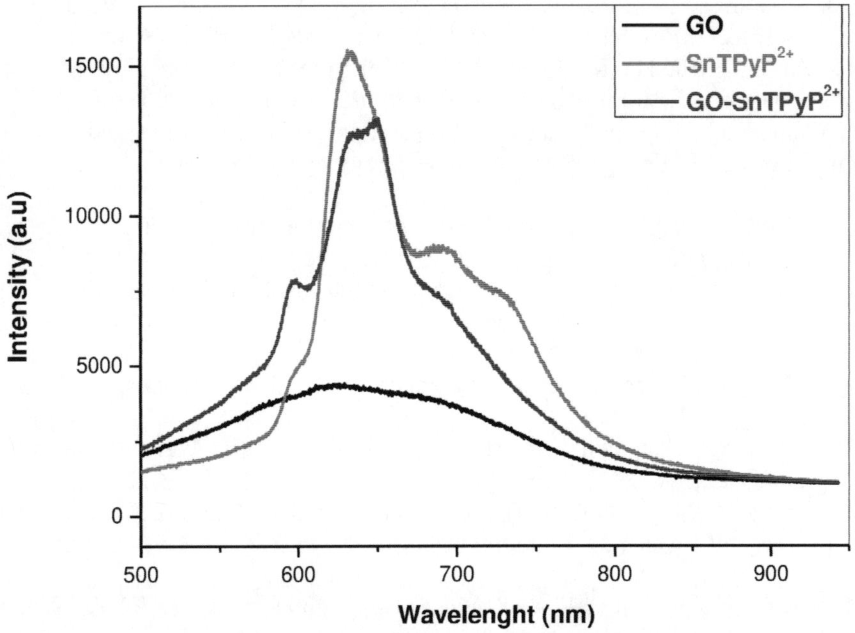

Figure 5. photoluminescence spectra of GO, $SnTPyP^{2+}$, and GO- $SnTPyP^{2+}$

These samples are obtained very simply by leaving a drop of solution to dry. We acquired a PL spectrum at regular intervals. Note after inserting GO in $SnTPyP^{2+}$ the intensity undergoes a decrease. In addition, the acquired spectrum of GO- $SnTPyP^{2+}$ composite has a low red shift and a slight narrowing relative to $SnTPyP^{2+}$ spectrum. The discrepancy is probably due to the change in the electronic structure of the composite, which shows that there is an interaction between GO and $SnTPyP^{2+}$ always gives rise to energy / charge transfer. Subsequently, the band around 590 nm characteristic of $SnTPyP^{2+}$ becomes clearer in the GO-$SnTPyP^{2+}$ composite. This illustrates the fact that the transfer involves quenching the luminescence of the porphyrin and the higher the luminescence of the donor is attenuated and the transfer is efficient. it is found that the extinction of the porphyrin is very important when it is in the presence of the GO. It shows a resonance at 580 nm corresponding to the S22 transition and provide evidence of a transfer of energy / charge of the molecule to the graphene oxide.

Conclusion

In this work we have synthesized graphene oxide-$SnTPyP^{2+}$ composite and studied the effect of concentration and pH on different optostructural properties. We demonstrated that the pH value is strongly related to the photon absorption in the composites. Vibrational and structural studies revealed that the composites we synthesized, present strong interaction between its components attributed to π-π interaction. The GO can thus serve as efficient electron acceptors, which opens the way to construct novel complexes with optimized pH and concentrations, with various types of porphyrins-graphene derivatives and other carbon nanomaterials.

IOP Conf. Series: Journal of Physics: Conf. Series **984** (2018) 012004 doi:10.1088/1742-6596/984/1/012004

References

[1] K.S. Novoselov, A.K. Geim, S.V. Morozov, D. Jiang, Y. Zhang, S.V. Dubonos, I.V. Grigorieva, A.A. Firsov, Electric Field Effect in Atomically Thin Carbon Films,2004 Science, **306** 666-669.
[2] Jianfeng Shen N , Min Shi, Yizhe Hu, Mingxin Ye, Covalent synthesis of organophilic chemically functionalized graphene sheets, 2010 Journal of Colloid and Interface Science, **348** 1-7.
[3] H. Chang, H. Wu, Graphene-based nanocomposites: preparation, functionalization, and energy and environmental applications, 2013 Energy & Environmental Science, **6** 3483-3507.

[4] D.A.C. Brownson, D.K. Kampouris, C.E. Banks, Graphene electrochemistry: fundamental concepts through to prominent applications, 2012 Chemical Society Reviews, **41** 6944-6976.
[5] Panda P , Ladomenou K , Coutsolelos A, Porphyrins in bio-inspired transformations: Light-harvesting to solar cell, 2010 J.Coordination Chemistry Reviews, **256** 2601-2627.
[6] W. Ai, Z. Du, Z. Fan, J. Jiang, Y. Wang, H. Zhang, L. Xie, W. Huang, T. Yu, Chemically engineered graphene oxide as high performance cathode materials for Li-ion batteries, 2014 Carbon, **76** 148-154.
[7] X.W. Xiaolin Li, Li Zhang, Sangwon Lee, Hongjie Dai, Chemically Derived, Ultrasmooth Graphene Nanoribbon Semiconductors, 2008 SCIENCE, **319** 5.
[8] Y. Liu, J. Zhou, X. Zhang, Z. Liu, X. Wan, J. Tian, T. Wang, Y. Chen, Synthesis, characterization and optical limiting property of covalently oligothiophene-functionalized graphene material, 2009 Carbon, **47** 3113-3121.
[9] S. Stankovich, D.A. Dikin, R.D. Piner, K.A. Kohlhaas, A. Kleinhammes, Y. Jia, Y. Wu, S.T. Nguyen, R.S. Ruoff, Synthesis of graphene-based nanosheets via chemical reduction of exfoliated graphite oxide, 2007 Carbon, **45** 1558-1565.
[10] H. Chang, H. Wu, Graphene-based nanocomposites: preparation, functionalization, and energy and environmental applications, 2013 Energy & Environmental Science, **6** 3483-3507.
[11] A.C. Ferrari, Raman spectroscopy of graphene and graphite: Disorder, electron–phonon coupling, doping and nonadiabatic effects, 2007 Solid State Communications, **143** 1-11.
[12] M.K. Panda, K. Ladomenou, A.G. Coutsolelos, Porphyrins in bio-inspired transformations: Light-harvesting to solar cell, 2012 Coordination Chemistry Reviews, **256** 2601-2627.
[13] S.K. N. Mongwaketsi, B. Klumperman, R. Sparrow , M. Maaza, Synthesis and characterization of porphyrin nanotubes/rods for solar radiation harvesting and solar cells, 2012 Physica B, **407** 1-5.
[14] D. Narayana Rao, Excited state dynamics in porphyrins in relevance to third-order nonlinearity and optical limiting, 2003 Optical Materials, **21** 45-49.
[15] S. FUKUZUMI, Electron transfer chemistry of metalloporphyrins and related metal complexes, 2000 Journal of Porphyrins and Phthalocyanines, **04** 398-400.
[16] Y.M. Akiharu Satake, and Yoshiaki Kobuke, Porphyrin-Carbon Nanotube Composites Formed by Noncovalent Polymer Wrapping, 2005 Chem. Mater., **17** 716-724.
[17] M.Y. Su Young Ryu, Jin Ho Choy, Sung Ho Hwang, Akihiro Frube, Tsuyoshi Asahi, and Hiroshi Masuhara, Optical Properties of Oxotitanium (IV) Meso-tetrakis(4-sulfonatophenyl)porphyrin Intercalated into the Layered Double Hydroxides (LDH) Studied by Laser Spectroscopy, Bull. Korean 2003 Chem. Soc., **24**.

[18] M. Khenfouch, J. Wéry, M. Baïtoul, M. Maaza, Photoluminescence and dynamics of excitation relaxation in graphene oxide-porphyrin nanorods composite, 2014 Journal of Luminescence, **145** 33-37.
[19] W. Chen, L. Yan, P.R. Bangal, Preparation of graphene by the rapid and mild thermal reduction of graphene oxide induced by microwaves, 2010 Carbon, **48** 1146-1152.
[20] A.C. Ferrari, Raman spectroscopy of graphene and graphite: Disorder, electron–phonon coupling, doping and nonadiabatic effects,2007 Solid State Communications, **143** 47-57.

[21] J.C.M. A. C. Ferrari, V. Scardaci, C. Casiraghi, M. Lazzeri, F.Mauri, S. Piscanec1, Da Jiang, K. S. Novoselov, S. Roth, and A. K. Geim, 2008 The Raman Fingerprint of Graphene,**5**.
[22] G. Wang, X. Shen, B. Wang, J. Yao, J. Park, Synthesis and characterisation of hydrophilic and organophilic graphene nanosheets, 2009 Carbon, **47** 1359-1364.
[23] A.C. Ferrari, J.C. Meyer, V. Scardaci, C. Casiraghi, M. Lazzeri, F. Mauri, S. Piscanec, D. Jiang, K.S. Novoselov, S. Roth, A.K. Geim, Raman spectrum of graphene and graphene layers, 2006 Phys Rev Lett, **97** 187401.
[24] J. Jiang, R. Pachter, F. Mehmood, A.E. Islam, B. Maruyama, J.J. Boeckl, A Raman spectroscopy signature for characterizing defective single-layer graphene: Defect-induced I(D)/I(D′) intensity ratio by theoretical analysis, 2015 Carbon, **90** 53-62.
[25] K. Krishnamoorthy, M. Veerapandian, K. Yun, S.J. Kim, The chemical and structural analysis of graphene oxide with different degrees of oxidation, 2013 Carbon, **53** 38-49.
[26] D.W. Lee, V.L. De Los Santos, J.W. Seo, L. Leon Felix, D.A. Bustamante, J.M. Cole, C.H. Barnes, The structure of graphite oxide: investigation of its surface chemical groups, 2010 The journal of physical chemistry. B, **114** 5723-5728.
[27] J.A.a.S.S. Choi, INFRARED SPECTROSCOPY OF PORPHYRINS, Annals New York 1973 Academy of Sciences, **206** 18.
[28] M. Aydin, DFT and Raman spectroscopy of porphyrin derivatives: Tetraphenylporphine (TPP), 2013 Vibrational Spectroscopy, **68**.
[29] Z.L. Mingshan Zhu, Bin Xiao, Yongtao Lu, Yukou Du, Ping Yang, and Xiaomei Wang, Surfactant Assistance in Improvement of Photocatalytic Hydrogen Production with the Porphyrin Noncovalently Functionalized Graphene Nanocomposite, 2013 ACS Appl. Mater., **5** 1-9.
[30] V.R.K. Satish Bykkam, Shilpa Chakra CH. and Tejaswi Thunugunta, SYNTHESIS AND CHARACTERIZATION OF GRAPHENE OXIDE AND ITS ANTIMICROBIAL ACTIVITY AGAINST Klebseilla AND Staphylococus, 2013 International Journal of Advanced Biotechnology and Research, **4** 1-5.
[31] M.B. Mohammed Khenfouch, Malik Maaza, White photoluminescence from a grown ZnO nanorods/graphene hybrid nanostructure, 2012 Optical Materials, **34** 1320–1326.
[32] H.D.P. Viet Hung Pham, Thanh Truong Dang, Seung Hyun Hur, Eui Jung Kim, Byung Seon Kong, Sunwook Kim and Jin Suk Chung, Chemical reduction of an aqueous suspension of graphene oxide by nascent hydrogen, 2012 The Royal Society of Chemistry.
[33] V. Štengl, S. Bakardjieva, M. Bakardjiev, B. Štíbr, M. Kormunda, Carborane functionalized graphene oxide, a precursor for conductive self-assembled monolayers,2014 Carbon, **67** 336-343.
[34] Y.-G.G. Jin-Song Hu, Han-Pu Liang, Li-Jun Wan, and Li Jiang, Three-Dimensional Self-Organization of Supramolecular Self-Assembled Porphyrin Hollow Hexagonal Nanoprisms, 2005 J. AM. CHEM. SOC., **127** 17090-17095.
[35] Y.C.a.J.J. Liang Wang, Controlling the growth of porphyrin based nanostructures for tuning third-order NLO properties, 2014 Nanoscale,, **6** 8.
[36] Song W, He C , Zhang W, Chen Y, Li X, Dong Y, Synthesis and nonlinear optical properties of reduced graphene oxide hybrid material covalently functionalized with zinc phthalocyanine,2014 Carbon, **77** 1020-1030.
[37] A. Kundu, R.K. Layek, A. Kuila, A.K. Nandi, Highly Fluorescent Graphene Oxide-Poly(vinyl alcohol) Hybrid: An Effective Material for Specific Au3+ Ion Sensors, 2012 ACS Applied Materials & Interfaces, **4** 5576-5582.
[38] H. Wu, W. Lu, J.-j. Shao, C. Zhang, M.-b. Wu, B.-h. Li, Q.-h. Yang, pH-dependent size, surface chemistry and electrochemical properties of graphene oxide, 2013 New Carbon Materials, **28** 327-335.
[39] G. Wang, B. Wang, J. Park, J. Yang, X. Shen, J. Yao, Synthesis of enhanced hydrophilic and hydrophobic graphene oxide nanosheets by a solvothermal method, 2009 Carbon, **47** 68-72.
[40] B. Minaev, M. Lindgren, Vibration and Fluorescence Spectra of Porphyrin-Cored 2,2-Bis(methylol)-propionic Acid Dendrimers, 2009 Sensors, **9** 1937-1966.

[41] T.-x. Ye, S.-l. Ye, D.-m. Chen, Q.-a. Chen, B. Qiu, X. Chen, Spectroscopic characterization of tetracationic porphyrins and their noncovalent functionalization with graphene, 2012 Spectrochimica Acta Part A: Molecular and Biomolecular Spectroscopy, **86** 467-471.

Synthesis and characterization of Graphene oxide/Zinc oxide nanorods sandwich structure

I Boukhoubza[1,3], M Khenfouch[2,3], M Achehboune[1,3], B Mouthudi[2], I Zorkani[1] and A Jorio[1]

[1] University Sidi Mohammed Ben Abdellah, Faculty of Sciences Dhar el Mahraz, Laboratory of Solid state Physics, Group of Nanomaterials and Renewable Energies, PO Box 1796 Atlas Fez 30 000, Morocco

[2] Department of Physics, College of Science, Engineering and Technology, Science Campus, University of South Africa, Cnr Christiaan de Wet & Pioneer Avenue Florida 1709, Johannesburg, South Africa

[3]Africa Graphene Center, Department of Physics, College of Science, Engineering and Technology, Science Campus, University of South Africa, Cnr Christiaan de Wet & Pioneer Avenue Florida 1709, Johannesburg, South Africa.

E-mail: boukhoubza.issam00@gmail.com

Abstract. Graphene-ZnO nanostructures composite materials have been used as very efficient candidates for various optoelectronic applications. Nowadays, the composite structure formation of ZnO nanostructures with graphene or graphene oxide is a novel, cost effective and efficient approach to control the morphology, surface defect states, band gap of ZnO nanocrystals .In this paper, we have prepared ZnO nanorods between two layers graphene oxide (GO/ZnO NRs/GO) via a simple hydrothermal method. Their morphology, structural and optical properties have been investigated. The obtained results of our composites GO/ZnO NRs/GO presented here showing an enhancement in the structural and optical properties. Thus may hold great promise to the development of the optoelectronic devices.

1. Introduction

Recently, Zinc oxide became one of the most promising materials for optoelectronic applications, gas sensing devices, transparent conductive oxide in solar cells, short wavelength electronic devices. For these applications, ZnO presents strong advantages: a wide direct band gap (3.4 eV), high electron mobility of 155 $cm^{-2}v^{-1}s^{-1}$, a high exciton binding energy (60 meV) at room temperature, low cost, and facility for synthesizing nanostructures with different morphologies [1]. One-dimensional ZnO nanostructures such as nanorods [2], nanowires [3], nanobelts [4], nanotube [5] and others have attracted attention owing to their interesting physical, chemical, optical, and electronic properties. A number of chemical and physical techniques have been used to easily synthesize ZnO nanostructures such as chemical bath deposition (CBD) [6], metal-organic chemical vapour deposition (MOCVD) [7], green synthesis method [8], solvothermal method [9], hydrothermal method [10] and microwave-assisted hydrothermal synthesis [11].

Graphene is a single monolayer of graphite, which has recently attracted much attention due to its unique electrical and optical properties. Furthermore, graphene has a very high carrier mobility and remarkable electroconductivity, mechanical flexibility, high chemical stability and optical transparency [12–15]. On the other hand, graphene oxide is receiving an increasing attention because it possesses the similar properties of graphene, as well as the special surface structures with the introduced hydroxyl and carboxyl groups for synthesis of graphene oxide containing nanocomposite [16,17]. Also, Graphene oxide contains a range of reactive oxygen functional groups, is considered to be a good candidate for chemical functionalization [18]. Thus, graphene oxide additives are often semiconducting, metallic or even polymeric in nature [19-21].

Composites based on graphene oxide and semiconductor nanostructures have attracted more and more attention of scientist researches, because graphene based composite materials shows potential as a class of new exciting materials [22], and it has large variety of current and future technological applications. A great number of inorganic nanostructures have been composited with graphene or graphene oxide, which include inorganic nanoparticles such as Fe_3O_4, TiO_2, SnO_2 and ZnO [23-26]. Currently, the composites formation of ZnO nanorods with graphene or graphene oxide is still novel and few studies discussing their elaboration have been reported in the literature. For example, Qin et al [27] developed a facile room temperature approach for graphene oxide-ZnO nanorods composites with enhanced photocatalytic activity. Alaver et al [28] reported the optical and structural properties of ZnO nanorods grown on graphene oxide and reduced graphene oxide by hydrothermal method. Vessali et al [29] synthesized ZnO NRs/ graphene oxide sheets using a chemical bath deposition. Their result shown that GO sheets deposited below the ZnO nanorods should affect the sensor resistance, reducing its responses to volatile organic compounds detection when compared to ZnO NRs sample.

In this study, the preparation of graphene oxide/ ZnO nanorods composites with sandwich structure via a simple hydrothermal method is reported. The optical, structural and morphology of graphene oxide/ZnO nanorods/graphene oxide were studied by X-ray diffraction (XRD), Scanning electron microscopy (SEM), Fourier transform infrared spectroscopy (FTIR) and UV-Vis spectroscopy. The obtained results reveal the efficiency of our methodology to improve the structural and optical properties of GO/ZnO NRs/GO nanocomposites, which might have great potential applications especially in optoelectronic devices

2. Experimental method

2.1. Synthesis of ZnO nanorods

ZnO precursors used in this work were purchased from Sigma-Aldrich. To synthesize ZnO nanorods, an equi-molar aqueous solution of 0.1 M zinc nitrate ($Zn(NO_3)_2$) and hexamethylenetetramine ($C_6H_{12}N_4$, HMT) was prepared using deionized water. This mixture was heated at a constant temperature of 90°C in an oven for 24 h. The obtained materials were deposited on Si ⟨1 0 0⟩ substrates and dried at 50°C for 2 hours.

2.2. Synthesis of graphene oxide(GO)

GO was prepared from graphite powder using modified hummers method [30]. Initially, 2 g of graphite powder was added to 46 ml of concentrated H_2SO_4 and kept in an ice bath for 2 h. Then, 6 g $KMnO_4$ was added gradually over a period of 30 min under stirring and cooling. Next, 46 ml deionized water was added slowly to this mixture and the temperature was maintained at ~100 °C for 15 min deionized water (280 ml) was then added, followed by 10 ml of 30% H_2O_2 solution, and the mixture was stirred for 30 min. The product was then filtered with 500 ml of 10% HCl solution to remove metal ions and then thoroughly washed with deionized water.

2.3. Synthesis of GO/ZnO NRs/GO composites
Firstly, a thin layers of GO was deposited on Silicon substrate and dried an oven at 60 °C vacuum for 1h. Secondly, ZnO nanorods were grown on the as-deposited GO surface. After 2h at 90 °C, another layer of GO covered the ZnO nanorods surface in order to prepare sandwich like structure composites.

3. Results and discussion

3.1. X-ray diffraction
The XRD pattern of Pure ZnO and GO/ZnO NRs/GO nanocomposites is shown in figure 1. The major diffraction peaks in GO/ZnO/GO composite were observed at 2θ value 31.8°, 34.5°, 36.5°, 47.8°, 56.6°, 62.8°, 66.5°, 67.9°, 69.3° and 76.9° which correspond to (100), (002), (101), (102), (110), (103), (200), (112), (201) and (202) crystalline plane of ZnO respectively. These crystalline planes are indexed to the hexagonal phase wurtzite structure of ZnO matched with the JCPDS No. 36-1451. The peaks at 2θ = 9.5° and 2θ = 23°corresponding to (002) plane of GO and reduced GO respectively [31-32]. The appearance of these two peaks is a consequence of the partial reduction of the GO which due to the presence of ZnO NRs between the layers of GO. Therefore, it can be clearly observed that the interaction between the components is remarkable as well as the intensity of (100) plane found to be higher than of the pure ZnO, which lead to a change in the orientation direction of ZnO NRs.

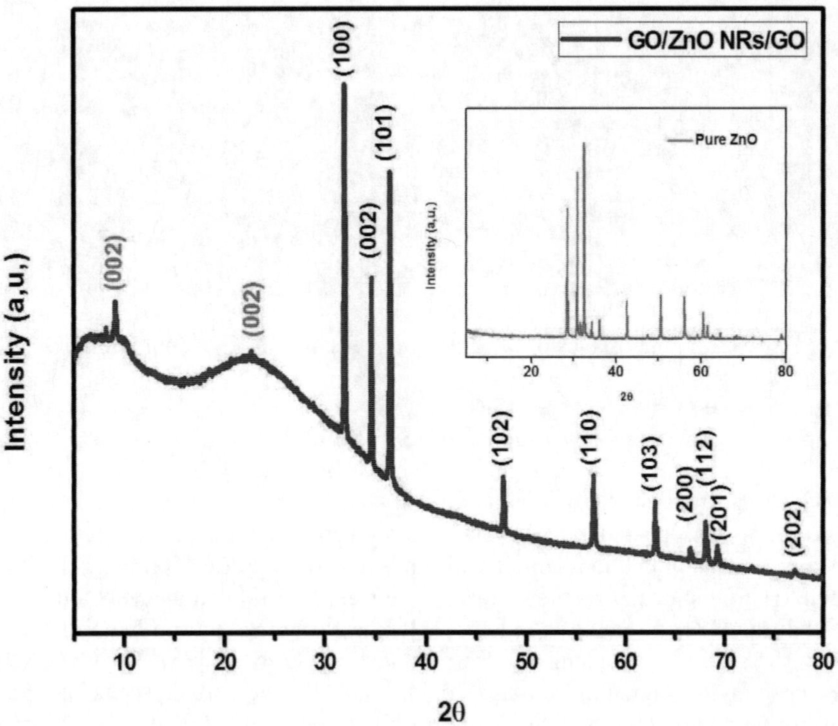

Figure 1. XRD pattern of GO/ZnO NRs/GO nanocomposites and Pure ZnO (Inset picture).

3.2. Scanning electron microscopy

The surface morphology of the ZnO nanorods and GO/ZnO NRs/GO composites were studied by SEM as illustrated in Figure 1. The as-prepared ZnO nanorods in figure 2(a) it can be seen that ZnO-NRs were grown vertically with high density and showed hexagonal shape. The average diameter of the nanorods was about 400 nm and the surface of ZnO nanorods was very smooth. For GO/ZnO NRs /GO sample shown in figure 2(b), it is clearly seen that ZnO NRs are covered by semi-transparent few layered GO sheets thus confirming the formation of sandwich-like nanocomposites structure by the insertion of the ZnO nanorods between the inter-layer of GO sheets which can prevent the agglomeration of ZnO nanorods as well as avoid the stacking of GO sheets and thus lead to preserve their high surface area.

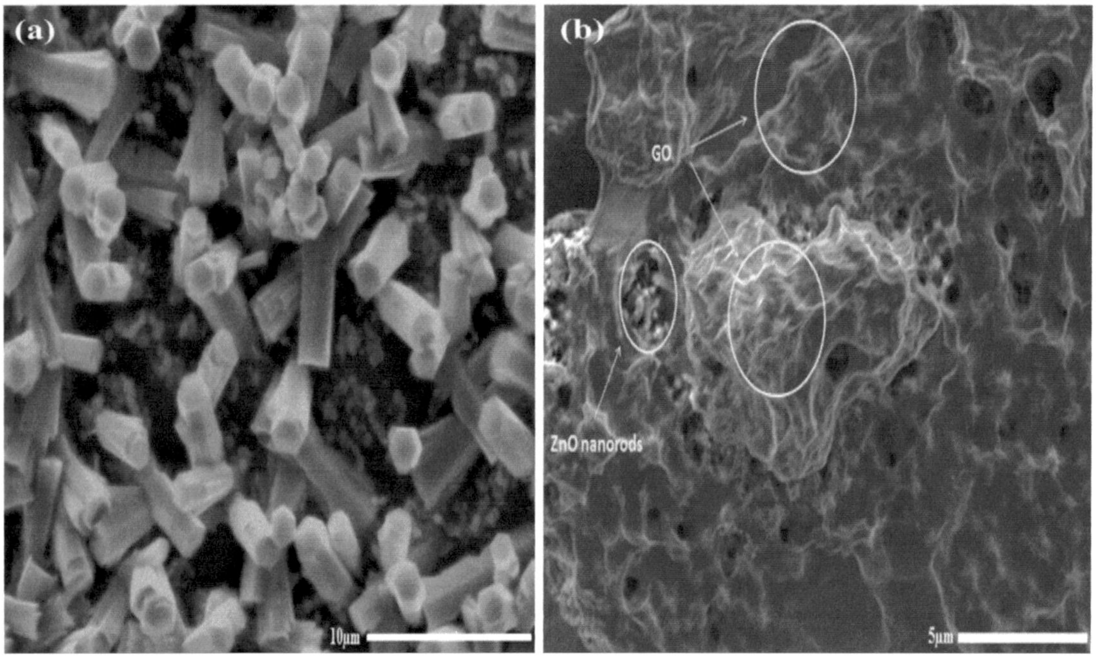

Figure 2. SEM Images of (a) ZnO nanorods and (b) GO/ZnO NRs/GO

3.3. Fourier transform infrared spectroscopy

The identification of molecular absorption surface of GO/ZnO NRs /GO nanocomposites was analyzed by FTIR spectroscopy technique and it is shown in Figure 3. The absorption band located at ~ 3340 cm^{-1} which can be assigned to the stretching vibrations of O–H groups from absorbed H_2O molecules, is almost disappeared due to the deoxygenation/reduction. The reduction makes the hydroxyl and epoxy edges of GO to interacts with ZnO fabrication [26]. Moreover, the peak at ~ 2347 cm^{-1} is assigned to CO_2 molecules on the surface of the samples [33] and the peak located at 2922 cm^{-1} is associated with the C-H stretching vibration [34]. The strong absorption band of the C-C skeletal vibrations around 1580 cm^{-1} which indicates the presence of GO sheets [35]. The characteristic peak at 1435 cm^{-1} corresponds to stretching vibration of aromatic C-OH and the peak at 957 cm^{-1} which corresponds to the stretching vibration of C-O becomes stronger [36]. Finally, the absorption peaks at 497 $cm^{-1}$and 565 cm^{-1} can be attributed to the stretching vibration of Zn-O. The region between 400-700 cm^{-1} usually corresponding to the vibration mode of metal-oxygen [37].

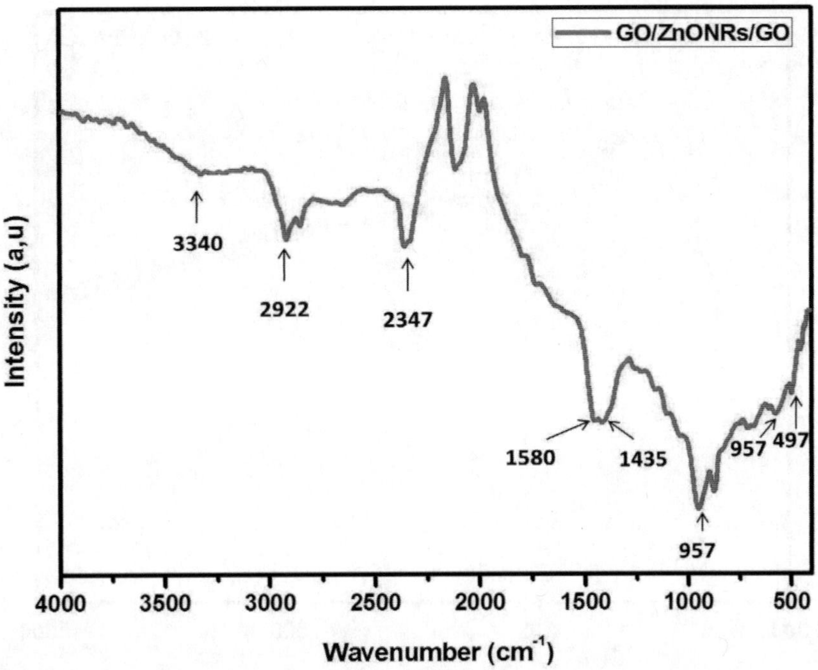

Figure 3. FTIR spectrum of GO/ZnO NRs/GO nanocomposites

3.4. UV-Vis spectroscopy

The ultraviolet-visible absorption spectrum of GO/ZnO NRs/GO nanocomposites was recorded to determine the effect of GO on the optical properties of ZnO nanorods and it is shown in Figure 4. The strong absorption band was observed at 370 nm corresponds to ZnO nanostructures. This indicates that the observed absorption peak is blue shifted from that of bulk ZnO due to nano size effect or may be attributed to the close conjugation of the ZnO nanorods and the graphene oxide that lead to high electron transfer which increased transition energy [31,38,39]. On the other hand the absoption peaks located at 370 and 412 nm can be attributed to $\pi \rightarrow \pi^*$ transition of aromatic C-C bands; from the valence to the conduction band ($O_{2p} \rightarrow Zn_{3d}$) as shown in the schematic diagram (figure 5) and the $n \rightarrow \pi^*$ transition of C=O bond respectively [40-41]. Moreover, we can mention also that the introduction of ZnO nanorods between two layers of GO has shown an enhancement in the light absorption by nanocomposites, which can be linked to the broad band absorption in the visible light region. However, our results show that the sandwich GO/ZnO NRs/GO structure is causing remarkable modifications on the optical character of ZnO NRs. Different transitions and absorption mechanisms indicated in the UV-Vis spectra were presented in the schematic diagram shown in the figure 5.

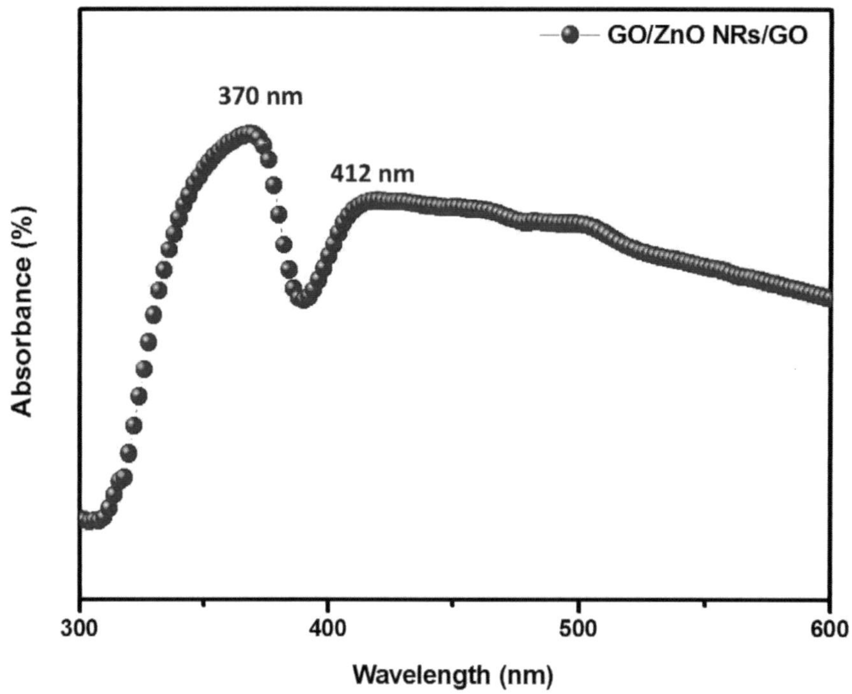

Figure 4. UV-Vis absorption spectrum of as-prepared GO/ZnO NRs/GO nanocomposites.

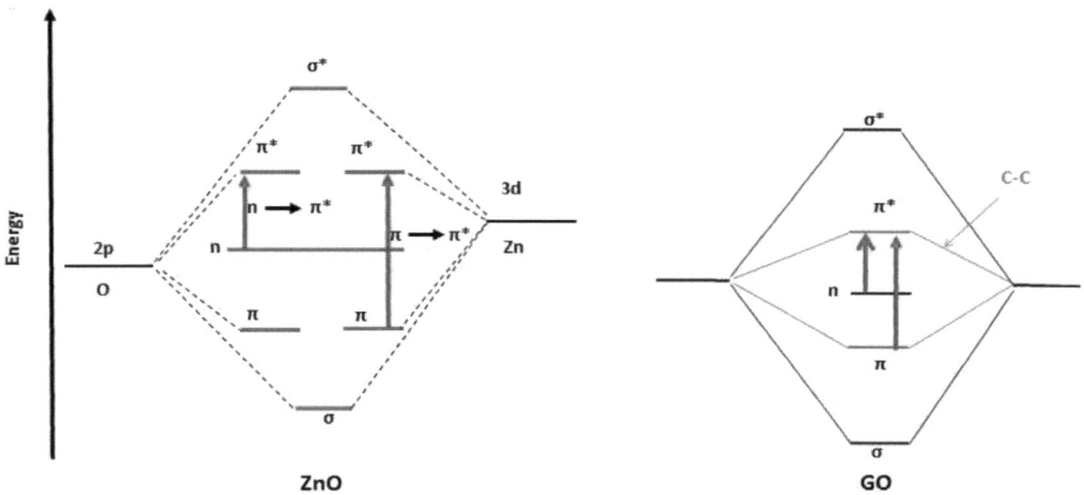

Figure 5. Schematic transitions diagram of ZnO and GO.

4. Conclusion

To summarize, GO/ZnO NRs/GO nanocomposites were synthesized via a simple hydrothermal synthesis method. The analysis of X-ray diffraction showed that the samples had hexagonal wurtzite structure, with (100) as a preferred orientation for the growth of the ZnO nanorods. SEM images also showed the formation of sandwich-like nanocomposites structure by the insertion of the ZnO nanorods between the inter-layer of GO while the vibrational properties are confirming the formation of these composites which assisted to the understanding of the growth mechanism. Furthermore, the GO/ZnO NRs/GO nanocomposites reveals a broad absorption in the whole region of (300-600 nm) where the strongest absorption was observed around 370 nm. The insertion of ZnO NRs between two layers of Go showed an enhancement of the optical properties, which can be linked to the broad absorption band in the visible light region. Therefore, these results can be used to enhance the efficiency of some optoelectronic applications.

Acknowledgments

This work was supported by the University Sidi Mohammed Ben Abdellah USMBA (Morocco), University of South Africa Department of Physics and Africa Graphene Center (South Africa).

References

[1] Hatamie A, Khan A, Golabi M, Turner A P F, Beni V, Mak W C, Sadollahkhani A, Alnoor H, Zargar B, Bano S, Nur N and Willander M 2015 *Langmuir* **31** 10913–10921
[2] Molaakbari E, Mostafavi A, Beitollahi H and Alizadeh R 2014 *Analyst* **139** 4356–4364
[3] Cui J 2012 *Materials Characterization* **64** 43-52
[4] H Asthana A, Yap Y K and Yassar R S 2015 *J. Nanosci. Nanotechnol* **15** 2277–2282
[5] Yu X, Song F, Zhai B, Zheng C and Wang Y 2013 *Physica E: low-Dimens Syst. Nanostruct.* **52** 92-96
[6] Poornajar M, Marashi P, Fatmehsari D H and Esfahani M K 2015 *Ceramics International* **42** 173-184
[7] Scalisi B A A, Toro R G, Malandrino G, Fragala M E and Pezzotti G 2008 *Chem. Vap. Deposition* **14** 115–122
[8] Mousavi M S S, Manteghi F, Kolahdouz M, Soleimanzadeh R, Norouzi M and Esfahan Z K 2015 Journal of *Alloys and Compounds* **650** *936-943*
[9] Rai P, Kwak W K and Yu Y T 2013 *Appl. Mater. Interfaces* **5** 3026–3032
[10] Yu J, Huang B, Qin X, Zhang X, Wang Z and Liu H 2011 *Appl. Surf. Sci.* **257** 5563-5565
[11] Rai P, Song H M, Kim Y S, Song M K, Oh P R, Yoon J M and Yu Y T 2012 *Mater. Lett.* **68** 90-93
[12] Cravotto G and Cintas P 2010 *Chem. Eur. J.* **16** 5246–5259
[13] KimK S, Zhao Y, Jang H, Lee S Y, Kim J M, Ahn J H, Kim P, Choi J Y and Hong B H 2009 *Nature* **457** 706–710.
[14] Eda G, Fanchini G and Chhowalla M 2008 *Nat. Nanotechnol.* **3** 270–274.
[15] Balandin A A, Ghosh S, Bao W, Calizo I, Teweldebrhan D, Miao F and Lau C N 2008 *Nano Lett.* **8** 902–907.
[16] Soci C, Zhang A, Xiang B, Dayeh S A, Aplin D P R, Park J, Bao X Y, Lo Y H and Wang D 2007 Nano Lett.**7** 1003–1009
[17] Boukhvalov D W and Katsnelson M I 2008 *Phys. Rev. B.* **78** 085413-085418
[18] Dreyer D R, Park S, Bielawski S W and Ruoff R S 2010 *Chem. Soc. Rev.* **39** 228
[19] Safa S, Sarraf-Mamoory R and Azimirad R 2015 *J Sol-Gel SciTechnol* **74** 499-506
[20] Sreeprasad T S, Maliyekkal S M, Lisha K P and Pradeep T, *J Hazardous Mater* 2011, **186** 921–931
[21] Mao Y, Wen S, Chen Y, Zhang F, Panine P, T.Chan T W, Zhang L, Liang Y and Liu L 2013 *Sci*

Rep **3** 2508 1-7
[22] Zou R, He G, Xu X, Liu Q, Zhang Z and Hu J 2013 *J. Mater. Chem. A* **1** 8445–8452
[23] Wang L, Jia X, Li Y, Yang F, Zhang L, Liu L, Ren X and Yang H *2014 J. Mater. Chem. A* **2** 14940-14946
[24] Kamat P V 2010 *J. Phys. Chem. Lett.* **1** 520–527
[25] Paek S M, Yoo E J and Honma I 2009 *Nano Lett.* **9** 72-75
[26] Kashinath L, Namratha K and Byrappa K 2015 *Applied Surface Science* **357** 1849-1856
[27] Qin J, Zhang X, Xue Y, Kittiwattanothai N, Kongsittikul P, Rodthongkum N, Limpanart S, Ma M and Liu R 2014 *Applied Surface Science* **321** 226-232
[28] Alver U, Zhou W, Belay A B, Krueger R, Davis K O and Hickman N S 2012 *Applied Surface Science* **258** 3109–3114
[29] Vessalli B A, Zito C A, Perfecto T M, Volanti D P and Mazon T 2017 *Journal of Alloys and Compounds* **696** 996-1003
[30] Thema F T, Moloto M J, Dikio E D, Nyangiwe N N, Kotsedi L, Maaza M and Khenfouch M 2013 *Journal of Chemistry* **1** 1-6
[31] Khenfouch M, Baïtoul M and Maaza M 2012 *optical Materials* **34** 1320–1326
[32] Rajeswari V, Jayavel R and Dhanemozhi A C 2017 *Materials Today: Proceedings* **4** 645–652
[33] Pandey P, Kurchania R, and Haque F Z 2014 *J. Adv. Phys.* 2014 **3** 1–7
[34] Zhengshan T, Suzhen B, Kesheng C and Jitao L 2016 *Mater. Express* **6** 437-443
[35] Sagadevan S, Pal K, Chowdhury Z Z and Foley M 2017 *Journal of Alloys and Compounds* **728** 645-654
[36] Huang K, Li Y H, Lin S, Liang C, Wang H, Ye C X, Wang Y J, Zhang R, Fan D Y, Yang H J,Wang Y G, Lei M 2014 *Powder Technology* **257** 113–119
[37] Panda N R, Sahu D, Acharya B S and Nayak P, 2015 *Current Applied Physics* **15** 389-396
[38] Beura R and Thangadurai P 2017 *Journal of Physics and Chemistry of Solids* **102** 168-177
[39] Sagadevan S, Pal S, Chowdhury Z Z and Foley M 2017 *Journal of Alloys and Compounds* **728** 645-654
[40] Khare P S, Yadav R and Swarup A 2013 International Journal of Applied Physics and Mathematics **3** 95-97
[41] Zak A K, Abrishami M E, Majid W H, Yousefi R, Hosseini S M 2011 *Ceram Inter* **37** 393–398.

Microwave assisted growth of nanorods vanadium dioxide VO_2 (R): structural and electrical properties

I Derkaoui[1,3], M Khenfouch[2,3], B M Mothudi[2], S J Moloi[2], I Zorkani[1], A Jorio[1] and M Maaza[4,5]

[1] University Sidi Mohammed Ben Abdellah, Faculty of Sciences Dhar el Mahraz, Laboratory of Solid state Physics, Group of Nanomaterials and Renewable Energies, PO Box 1796 Atlas Fez 30 000, Morocco

[2] Department of Physics, University of South Africa, Private Bag X90, Florida, 1710, South Africa

[3] Africa Graphene Center, Physics department, Eureka building, College of Science, Engineering and Technology, Science Campus, University of South Africa, Cnr Christiaan de Wet & Pioneer Avenue Florida 1709, Johannesburg, South Africa

[4] Nanosciences African Network (NANOAFNET), iThemba LABS-National Research Foundation, 1 Old Faure road, Somerset West 7129, PO Box 722, Somerset West, Western Cape-South Africa

[5] UNESCO-UNISA Africa Chair in Nanosciences Nanotechnology, College of Graduate Studies, University of South Africa, Muckleneuk ridge, PO Box 392, Pretoria-South Africa

Email: derkaouiissam@gmail.com

Abstract. Nanostructured metal oxides have attracted a lot of attention recently owning to their unique structural advantages and demonstrated promising chemical and physical properties for various applications. In this study, we report the structural and electrical properties of vanadium dioxide VO_2 (R) prepared via a single reaction microwave (SRC) synthesis. Our results are revealing that the components of VO_2 (R) films have a rod-like shape with a uniform size distribution. The nanorods with very smooth and flat surfaces have a typical length of up to 2µm and a width of about several nanometers. The structural investigations reveal the high crystallinity of VO_2 (R) ensuring good electrical contact and showing a high conductivity as a function of temperature. This synthesis method provides a new simple route to fabricate one-dimensional nanostructured metal oxides which is suitable for a large field of applications especially for smart windows.

1. Introduction

Among the transition metal oxides, nanostructured vanadium oxides materials, such as V_2O_5, V_2O_3, V_3O_7, VO_2 (B), VO_2 (M), VO_2 (R) and VO_2 (A) have been used in many fields due to their outstanding structural flexibility [1] and their wide range of oxidation states (from +2 to +5) in the vanadium-oxygen system [2]. Vanadium dioxide (VO_2) is the most well-known of many vanadium-related oxides and has been extensively investigated for its distinctive multifunctional including chemical, optoelectronic and electrochromic properties [3]. VO_2 nanoparticles have been brought into sharp focus with its different crystalline phases [4]. Because of these attractive properties they exhibit potential applications in many areas such as smart windows [5] and energy storage devices [6].

Vanadium dioxide with rutile-type structure VO_2 (R) undergoes first-order reversible metal-insulator transition (MIT) at around 68 °C. During the phase transition process, there is a structural phase transformation from high-temperature metallic tetragonal phase VO_2 (R) to low-temperature insulating

monoclinic phase VO_2 (M_1) [7]. The reversible transition in structure is accompanied by reversible changes in its optical and electrical properties [8]. Many different synthesis methods have been developed for the synthesis of vanadium oxide nanoparticles including sol–gel [9], vapor transport [10], pulsed laser ablation [11] and hydrothermal method [12].
There are few reports devoted to the synthesis of VO_2 (R) by microwave method. The microwave route for synthesizing vanadium oxide nanoparticles is particularly successful in terms of controlling every reaction by direct pressure and temperature control. with very high growth rate and small particle sizes as a consequence of fast homogenous nucleation [13]. To the best of our knowledge, this is the first study of the effect of the time and annealing treatment on the growth of the nanostructured vanadium dioxide VO_2 (R) prepared by microwave SRC.
In the present work, we investigate the structural and electrical properties of vanadium dioxide VO_2 (R). The structure and morphology of our nanoparticles are investigated in order to probe the effects of the time and annealing treatment on the morphology and size change of the nanostructured vanadium oxides. In this sense, X-ray diffraction (XRD), transmission electron microscopy (TEM) and electrical conductivity were investigated.

2. Experimental

2.1. Material preparation

Vanadium oxides nanoparticles VO_2 (R) were synthesized using a single reaction microwave synthesis.

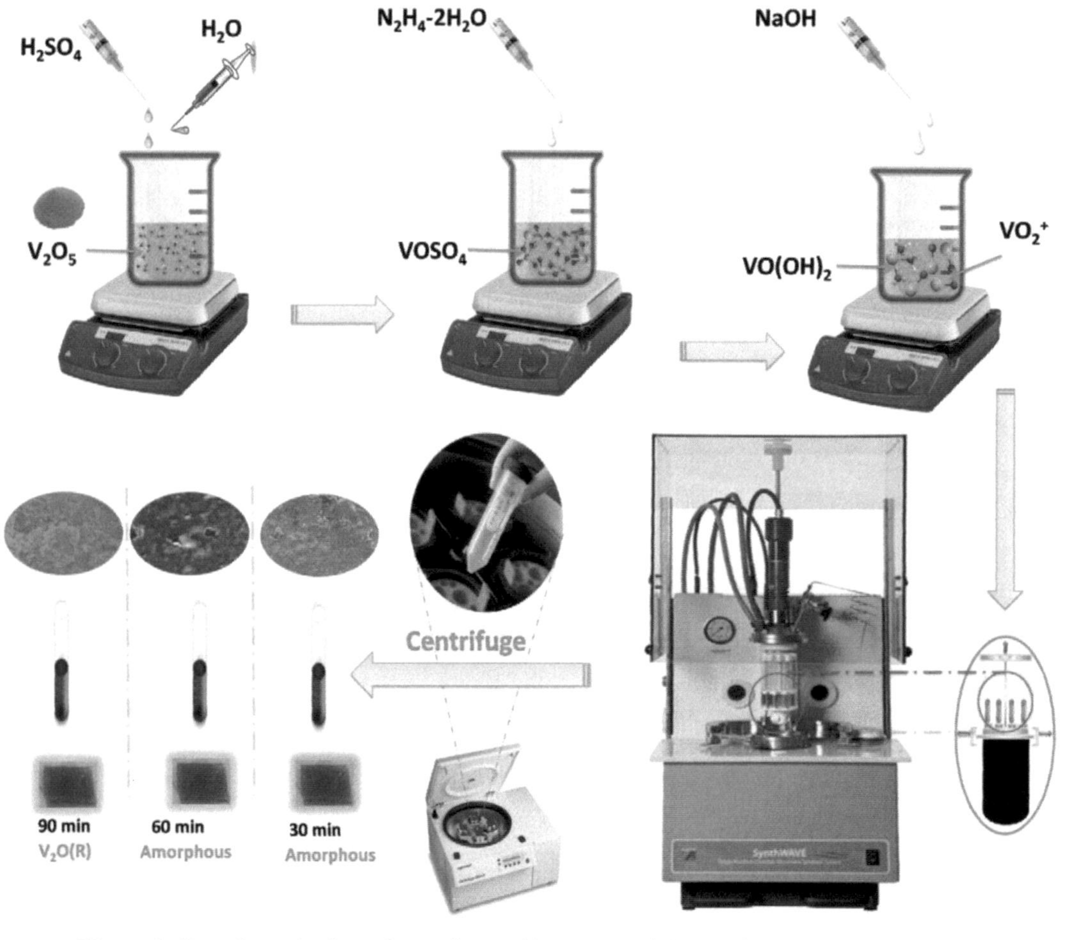

Figure 1. Growth mechanism of vanadium oxides nanoparticles synthesized by a single reaction microwave synthesis.

In a typical synthesis, V_2O_5 (0.9g) was first stirred with 20 ml of deionized water to make a yellow suspension. Then 1.5 ml of sulphuric acid, H_2SO_4, was added to the suspension while heating at around 60 °C. 0.25 ml of hydrazine hydrate, N_2H_4-$2H_2O$, was added dropwise to the solution. Heat could also be applied to the solution while adding NaOH and hydrazine solution. The pH of the blue VO^{2+} solution was strongly acidic, then was weighed with a stir bar and placed on a rack. The rack is fitted to the chamber top with a mechanical stirrer, which is lowered automatically. The mixture was treated at 230 ° C for 30, 60 and 90 minutes. When the program stops, a circulating water-cooling jacket rapidly cools the chamber, pressure is gently released, and gases and vapors are directed away to exhaust. The resulting suspensions were collected by centrifugation at 4500 rpm for 40 min to remove metal ions and impurities using distilled water and ethanol (figure 1). All of the chemical reagents were analytical grade. They were purchased from Acros Organics and used without further purification.

2.2. Material Characterization
To characterize the obtained samples, XRD patterns were obtained from a Rigaku Smart Lab system using Cu Kα (λ=1.54178 A°), scanning electron microscopy (SEM) is employed to observe the morphology of vanadium oxides nanoparticles. Finally, conductivity was measured with a Picoammeter/Voltage Source-Keithley Model 6487 system.

3. Results and discussions

3.1. X-ray diffraction

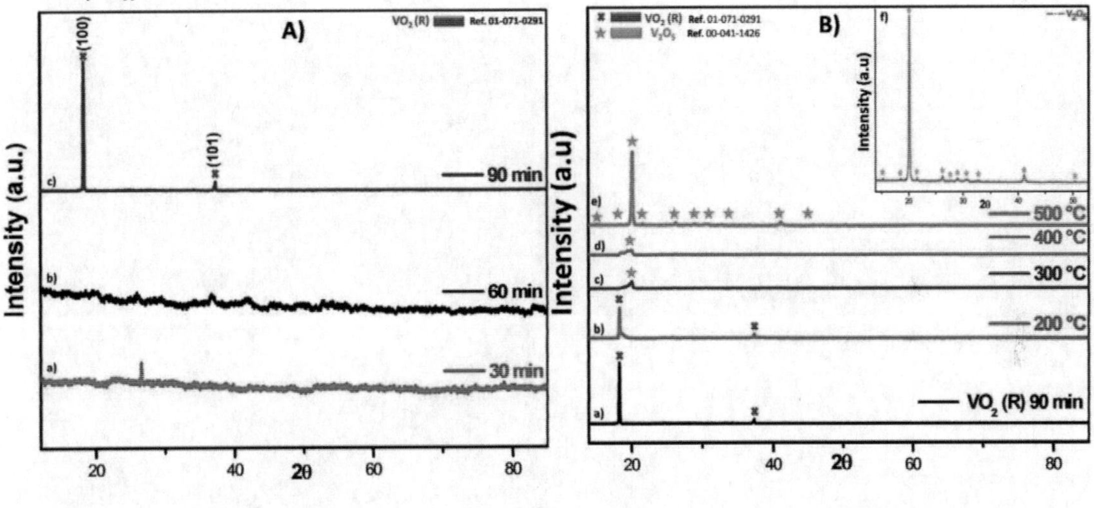

Figure 2. XRD patterns of VO-NPs (a) synthesized at different time and (b) under annealing process for 3h at 90 min.

The XRD pattern of vanadium oxides nanoparticles synthesized at different time (figure 2 .(A). (a) - (b)) revealed an amorphous phases, this indicates that the vanadium oxide nanoparticles are not efficiently formed during the growth mechanism. However, after 90 minutes, all the peaks could be indexed as the tetragonal VO_2 (R) (JCPDS: 71-0291) with lattice constants: a = b = 4.5546 A°, c = 2.8528 A° and β = 90.00°. There are only the neat and narrow peaks without any outsider peaks or impurity, revealing that the V^{5+} ions in V_2O_5 have been reduced to V^{4+} ions by the hydrazine in the reaction. Figure 2 (B) shows the XRD pattern of VO_2 (R) under annealing process for 3h at 200, 300, 400 and 500 °C respectively. We observe that we have similar phase of monoclinic VO_2 (R) at 200 °C for 3h accompanied by a decrease in the intensity of XRD spectrum. However, for 300 °C the peaks of VO_2 (R) vanished which indicated that the phase was modified with the appearance of vanadium pentoxide picks. In addition, at 500 °C, all the peaks could be indexed as the orthorhombic V_2O_5 (Space group 59: Pmmn) with lattice constants a = 11.5160 A°, b = 3.5656A° and c = 4.3727 A°, and

β = 90.00°; this is compatible with the standard value of JCPDS: 41-1426 already described in the literature [14]. Hence, Vanadium oxides can be selectively synthesized with controllable phase structures VO_2 (R) and V_2O_5 with increasing synthesis time and annealing temperature which affect the stoichiometry and the arranged band.

3.2. SEM Measurements

The SEM images of our nanoparticles synthesized at different time (figure 3 (a) - (b)) revealed that vanadium oxides nanoparticles were not efficiently formed during growth mechanism indicating that the phase was amorphous even after 60 min. The SEM image in figure 3 (c) confirms rod-shaped morphology of VO_2 (R) nanoparticles and shows an important modification on the morphology of these nanostructures prepared at 90 min. The average diameter of the nanorods was from several tens to several hundred nanometers and the surface was very smooth. These results confirm that time is a principal parameter in the formation of VO-NPs.

Furthermore, figure 3 (d) confirms that the morphology was changed after annealing treatment, it can be observed that the transition of crystal phases from VO_2 (R) to V_2O_5 in our experiments occurred at 500 °C. The growth of VO-NPs, in this case, leads to a sphere-like shape with very smooth and flat surfaces and a typical width of about several nanometers.

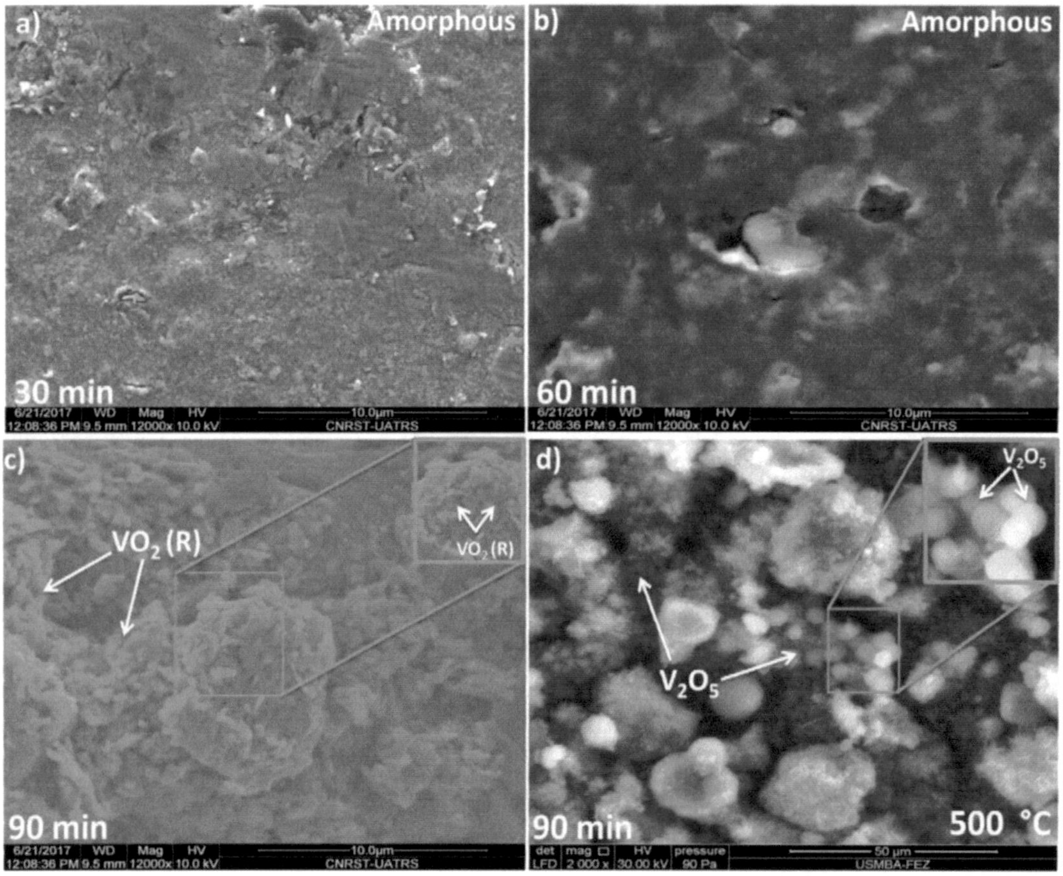

Figure 3. SEM surface morphology: (a) amorphous phase at 30 min, (b) amorphous phase at 60 min (c) VO_2 (R) nanorods at 90 min, (d) V_2O_5 nanospheres at 500 °C.

These results support the structural investigations and confirm the remarkable effect of time and annealing treatment on the growth of vanadium oxide nanostructures.

3.3. I-V Characterization

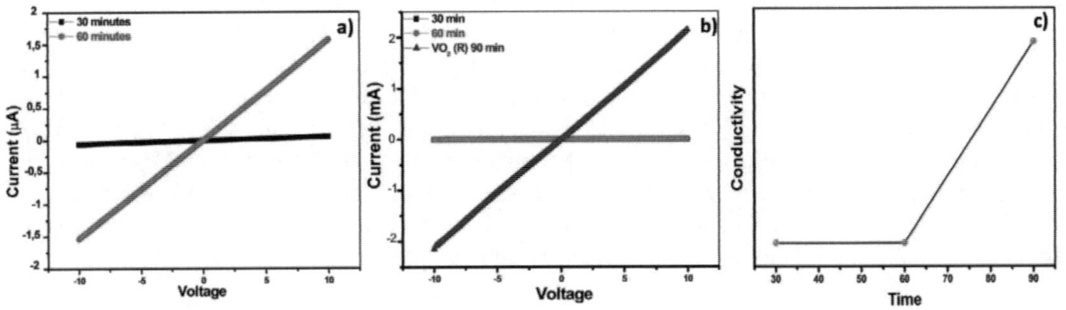

Figure 4. Conductivity measurement of VO-NPs nanostructures synthesized at different time: 30, 60 min (a, b) and the conductivity as a function of the time at the applied voltage (10 V) (c).

The I-V characteristics of VO-NPS nanostructures at different time are shown in figure 4. The representation of the I-V curves on the same figure makes it possible to better understand the influence of time on the electrical properties. As we can see (figure 4 (a) -(b)), the I-V curves of the VO-NPs obtained are linear indicating an ohmic behavior of the contact. At a negative voltage, for a period of 90 minutes, the value of the current at which the conductivity starts its linear behavior is -2 mA.
Figure 4 (c) shows the conductivity as a function of the time at the applied voltage (10 V). It is observed that from 30 to 60 min there is no a great enhancement in the conductivity. However, the conductivity increases rapidly with an increase of the time from 60 to 90 min The increment in the time reduces the occupancy of defect states, which is due to a large number of charge carriers available for conduction so that the mobility increases. Hence, this results shows that by increasing the time, the electronic conductivity can be increased by 3 orders of magnitude.

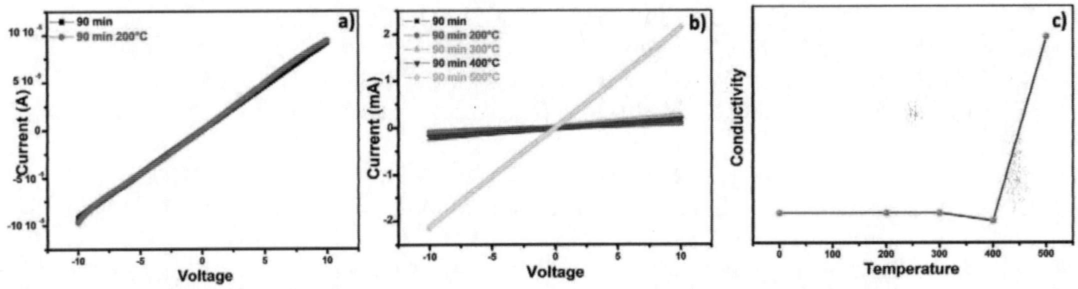

Figure 5. Conductivity measurement of VO-NPs nanostructures under annealing process for 3h at 90 min (a, b) and the conductivity as a function of the annealing treatment at the applied voltage (10 V) (c).

Figure 5 provides a comparison of the I-V characteristics of VO-NPs before and after the annealing treatment at 90 min. Earlier measurements show that by increasing the temperature, the electronic conductivity can be increased by 2 orders of magnitude [15-16]. In figure 5 (a)-(b), the measurement show that the conductivity increases with increasing temperature where the number of electronic defects decreases so the mobility increases. Figure 5 (c) shows the conductivity as a function of the temperature at the applied voltage (10 V). It is observed that the conductivity increases moderately with an increase of the annealing treatment from 0 °C to 300 °C. Then, we observed that above 400 °C the conductivity decreases slightly which may be due the change of the phases and morphology of VO-NPs [17]. However, for 400 to 500 °C we observed, that a large number of charge carriers available for conduction are accompanied by a greater increase of electrical conductivity. Our results show that the two main reasons for good electrical conductivity of VO-NP films are the increase in the time and annealing treatment.

4. Conclusion

The purpose of this study was to investigate the structural and electrical properties of vanadium oxides prepared by a single reaction microwave synthesis. The experimental results demonstrate that synthesis time and annealing treatment effecting on the growth of the nanostructured vanadium dioxide VO_2 (R). Our result highlighted the modification of VO-NPs form, shape and phase after the time and annealing treatment showing that the annealing and time are two main factors in controlling the phase and morphology of vanadium oxides nanoparticles. The experimental measurement revealing that the annealing state and the increase in time are two crucial parameters for the improvement of the electrical conductivity. The use of these nanoparticles is promising way for the development of technological applications for smart windows as solar heat control.

Acknowledgments

Special thanks to the Innovation city of USMBA (Morocco), Abdus Salam International Centre for Theoretical Physics (Trieste-Italy), Nanosciences African Network, University of South Africa Department of Physics, iThemba Labs (South Africa), African Laser Center and Africa Graphene Center.

References

[1] Kudo T, Ikeda Y, Watanabe T, Hibino M, Miyayama M, Abe H and Kajita K 2002 *Solid. State. Ionics.* **152** 833
[2] Jacques L 2010 *Mater.* **3** 4175
[3] Parida M R, Vijayan C, Rout C S, Sandeep C S, Philip R and Deshmukh P C 2010 *J. Phys. Chem.* C **115** 112
[4] Cao C, Gao Y and Luo H 2008 *J. Phys. Chem.* C **112** 18810
[5] Balu R and Ashrit P V 2008 *Appl. Phys. Lett.* **92** 021904
[6] Huanwen W, Huan Y, Xiao C and Xuefeng W 2014 *J. Mater. Chem.* A **2** 1165
[7] Son J H, Wei J, Cobden D, Cao G and Xia Y 2010 *Chem. Mater.* **22** 3043
[8] Zylbersztejn A and Mott N F 1975 *Phys. Rev.* B **11** 4383
[9] Mai L Q, Hu B, Hu T, Chen W and Gu E D 2006 *J. Phys. Chem.* B **110** 19083
[10] Sohn J I, Joo H J, Porter A E, Choi C J, Kim K, Kang D J and Welland M E 2007 *Nano. Lett.* **7** 1570
[11] Rozen J, Lopez R, Haglund R F and Feldman L C 2006 *Appl. Phys. Lett.* **88** 081902
[12] Derkaoui I, Khenfouch M, Elmokri I, Moloi S J, Mothudi B M, Dhlamini M S, Maaza M, Zorkani I and Jorio A 2016 *Graphene* **5** 14
[13] Jansen J C, Arafat A, Barakat A K and Bekkum H 1992 Microwave synthesis of nanocrystalline metal sulfides in formaldehyde solution *Synthesis of Microporous Materials* (*Occelli ML, Robson HE vol 1*) ed (New York, NY, USA: Van Nostrand Reinhold) chapter 33. pp. 507–521
[14] Liu H M and Yang W.S 2011 *Energy Environ. Sci.* **4** 4000
[15] El-Desoky M M 2002 *Mater. Chem. Phys.* **73** 259
[16] Montani R A, Lorente A and Vincenzo M A 2000 *Solid State Ionics* **130** 91
[17] Derkaoui I, Khenfouch M, Mothudi B M, Jorio A, Zorkani I and Maaza M 2017 *J. Mater. Sci. Mater. Electron.* **28** 17710

Structural and optical characterization of Holmium coated ZnO nanorods

M Achehboune[1,2], M Khenfouch[2,3], I Boukhoubza[1,2], B Mothudi[3], I Zorkani[1] and A Jorio[1]

[1]University Sidi Mohammed Ben Abdellah, Faculty of Sciences Dhar el Mahraz, Laboratory of Solid state Physics, Group of Nanomaterials and Renewable Energies, PO Box 1796 Atlas Fez 30 000, Morocco.

[2]Africa Graphene Center, Physics department, Eureka building, College of Science, Engineering and Technology, Science Campus, University of South Africa, Cnr Christiaan de Wet & Pioneer Avenue Florida 1709, Johannesburg, South Africa.

[3]Physics department, Eureka building, College of Science, Engineering and Technology, Science Campus, University of South Africa, Cnr Christiaan de Wet & Pioneer Avenue Florida 1709, Johannesburg, South Africa.

E-mail: achehboune.mohamed01@gmail.com

Abstract. One-dimensional ZnO nanostructures have been attracting growing research effort worldwide over the last few years due to their various morphologies, easy synthesis and excellent physical properties for fabricating optoelectronic devices. In this study, we have reported the synthesis of Holmium coated ZnO nanostructures. The structural, morphological and optical properties of our materials were studied using different techniques. The obtained materials show promising properties and the remarkable effects of the coated material. The experimental results obtained for our material exhibit signficant enhancement in properties, which is suitable for the optoelectronic applications.

1. Introduction

Zinc Oxide (ZnO) is considered as a very promising semi-conductor for optoelectronic and photonic applications such as solar cells, light emitting diodes, laser diodes and photodetectors due to its wide band gap (3.37eV), large exciton binding (60meV) and high electron mobility [1-3]. Pure ZnO has transmittance of more than 75% and low absorption and scattering values at high wavelengths, making it transparent over the visible range [4]. The variety of nanostructures of ZnO makes it as an attractive and new material with potential applications in many fields of nanotechnology. One-dimensional structures are considered the largest group of ZnO nanostructures including nanorods [5], nanowires [6], nanobelts [7] and nanotubes [8]. Various methods have been developed during several years for the synthesis of one-dimensional ZnO that have included hydrothermal method [9], chemical vapor deposition [10], green synthesis [11], sol-gel method [12] and electrochemical technique [13]. However, the hydrothermal method is very simple, easy and low-cost method used for obtaining nanostructures in a specific shape and size at lower temperature (lower than 100°C) [14].

Optoelectronic properties of ZnO nanostructures can be modified and improved by doping and coating processes with other materials; As known ZnO have been heavily doped and co-doped with metals such as Al [15], Ga [16], and Fe [17]. Therefore, rare earth doped materials exhibit very good luminescent properties owing to their 4f intra shell transition [18]. Among the Rare Earth (RE), holmium (Ho) is a promising element for applications with good luminescent properties because it exhibits fluorescence in the visible and near infrared region [19]. In the recent years, ZnO nanostructures have been doped by Rare earths (Ytterbium (Yb), Erbium(Er) and Holmium (Ho)); the study conducted by Zamiri et al [20] reported the synthesis of rare earth (Er, La, Yb) doped ZnO nanostructures through wet precipitation method and studied their structural and optical properties. Senol [21] have also studied the structural, optical and electrical properties of rare earth (Er, Yb)-doped ZnO using hydrothermal method. The study was carried out by Soumahoro et al [22] have reported the structural, optical and electrical properties of Yb-doped ZnO thin film synthesized by the spray pyrolysis method in order to investigate the effect of Yb ion in the ZnO matrix and the related optical properties of the films. Khataee et al [23] synthesized Holmium doped ZnO nanoparticles using sonochemical technique. Singh et al [24] have synthesized Ho and Y-doped ZnO nanoparticles using wet chemical synthesis technique, the optical studies reveal that the near band edge position shifts towards the lower wavelength side for the as synthesized nanoparticles. Franco et al [25] have also reported the optical and dielectric properties of Ho doped ZnO synthesized by the combustion reaction method.

Coating ZnO nanostructures could be one of the most effective approaches, which used to improve its properties. In spite of that, the studies on the coated ZnO nanostructures are relatively poor in literature; Li et al [26] have investigated the micro-structure and optical performance of Ag-coated ZnO nano-needles. Chen et al [27] have reported the characterization of freestanding graphene coated ZnO nanowires for optical waveguiding. To the best of our knowledge, the synthesis and characterization of ZnO nanorods coated with holmium have not been previously reported in literature.

In this present work, we report the synthesis and the structural characterization of pure and Holmium coated ZnO nanorods; hydrothermal method has been applied to synthesize uncoated and Ho-coated ZnO nanorods. X-ray diffraction (XRD), Scanning electron microscopy (SEM) and Fourier-transfer infrared spectroscopy (FTIR) were used to characterize the structural modifications induced by Holmium into the surface of ZnO and to survey the change in the morphology of samples. UV-visible spectroscopy was performed to study the absorbance properties of Ho coated ZnO nanorods. Hence, the fundamental aim of this current work is to investigate the effect of holmium on the structural and optical properties of ZnO nanorods.

2. Experimental details

In the hydrothermal process, ZnO precursors were purchased from Sigma-Aldrich. To synthesize ZnO nanorods, an equi-molar aqueous solution of 0.1 M zinc nitrate ($Zn(NO_3)_2$) and hexamethylenetetramine ($C_6H_{12}N_4$, HMT) was prepared using deionized water. This mixture was heated at a constant temperature of 90°C in an oven for 24 h. The coated material followed the same steps, then the surface of ZnO nanorods was coated with holmium using the drop coating.

The obtained materials were deposited on Si (100) substrates and dried at 50°C for 2 hours.

3. Results and discussion

3.1. X-Ray diffraction

Figure 1 shows the X-ray diffraction pattern of pure and Ho coated ZnO nanorods. As can be seen from the spectra, the analysis of the diffraction peaks of pure ZnO reveals the presence of hexagonal wurtzite structure with no other peaks of impurities. Moreover, it shows that the growth was in the three principal directions (reflections (100), (002), (101)). For Ho coated ZnO, results showed that (002) and (101) peaks of ZnO have been disappeared, while the corresponding diffraction peak (100) of ZnO still exist, but its intensity have been decreased. Intense diffraction peaks corresponding to holmium emerged at

$2\theta = 8.32°$, $2\theta = 9.38°$ and $2\theta = 18.51°$. This indicates that the ZnO nanorods were coated with Ho by using drop coating technique.

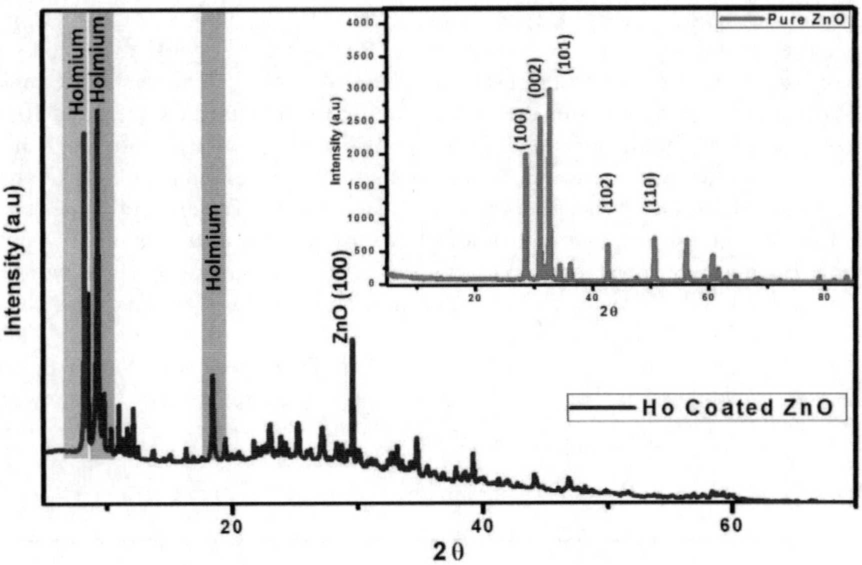

Figure 1. X-ray diffraction spectrum of pure and Ho coated ZnO nanorods.

3.2. Scanning electron microscopy

Morphological characterization of pure and Ho coated ZnO nanorods was studied using scanning electron microscopy (SEM) observation, Figure 2 illustrates typical SEM of ZnO nanorods prepared via hydrothermal method. As clearly shown in the Figure 2(a), pure ZnO nanorods were relatively grown vertically, nanorods having hexagonal shape with average diameter and length are about 500 nm and several micrometres (~4μm) respectively. For the coated material shown in the figure (b), ZnO nanorods grown with different sizes and different diameters. In addition, it is clear to see that the nanorods are coated tightly by the holmium with a few wrinkles and higher defect density on surface of nanorods, which indicates that the surface morphology of ZnO nanorods is influenced by coating with holmium.

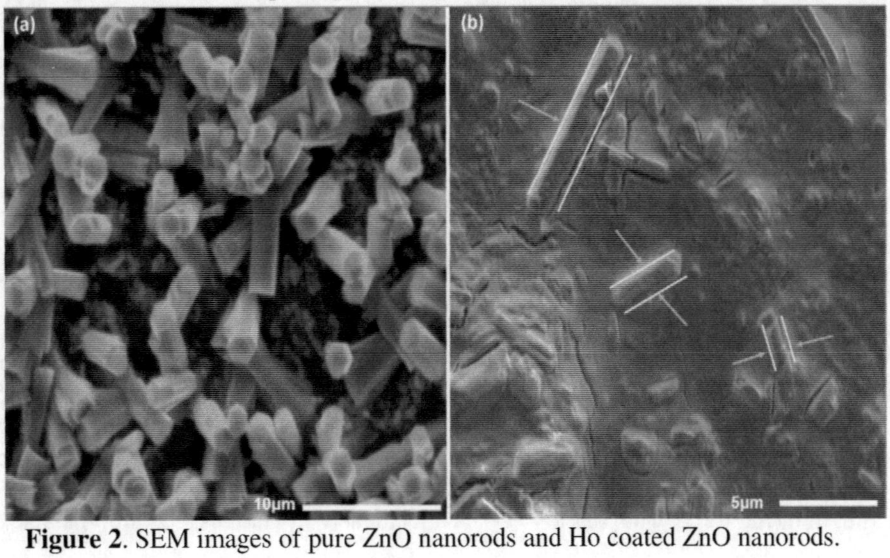

Figure 2. SEM images of pure ZnO nanorods and Ho coated ZnO nanorods.

IOP Conf. Series: Journal of Physics: Conf. Series **984** (2018) 012007 doi:10.1088/1742-6596/984/1/012007

3.3. Fourier transform infrared (FTIR) spectroscopy

Fourier transform infrared (FTIR) spectroscopy is an important technique, which reveals information about the structure orientation; chemical compositions of a compound and identification of impurities mainly exist near samples surface. ZnO coated with holmium were characterized by FTIR spectroscopy shown in the Figure 3. The coated material exhibits characteristic absorption peaks at 475 cm^{-1} and 674 cm^{-1} which could be assigned to Metal-Oxygen (Zn-O or Ho-O) stretching vibration. The peak located at 674 cm^{-1} is more intense which is may be correspond to Zn-O bending vibration mode of ZnO; the typical characteristic band of the wurtzite hexagonal phase of pure ZnO, normally the region band of 400-700 cm^{-1} corresponds to the metal-oxygen band [28]. The band observed at around 1050 cm^{-1} may be ascribed to the bending vibration of nitrate [29]. The two peaks located at 1460 cm^{-1} and 1660 cm^{-1} that may be assigned to the symmetric and asymmetrical stretching of Zinc carboxylate respectively; probably from the small residue of zinc acetate used in reaction [30]. The absorption peak around 2920 cm^{-1} is due to C-H (acetate) stretching mode of alkane groups and the peak around 3734 cm^{-1} is due to O-H stretching vibrations of hydroxyl group on the surface of the material [31]. However, the origin vibration of the strongest peak located at 2350 cm^{-1} can be assigned to adsorbed CO_2 molecules on the surface of the sample [32]. Comparing our measurements with the results of pure ZnO nanorods synthesized via hydrothermal method reported by Soni et al [30], there is a shift of the all bands in addition to the presence of strongest band at 2350 cm^{-1} indicating a strong interaction between ZnO and holmium.

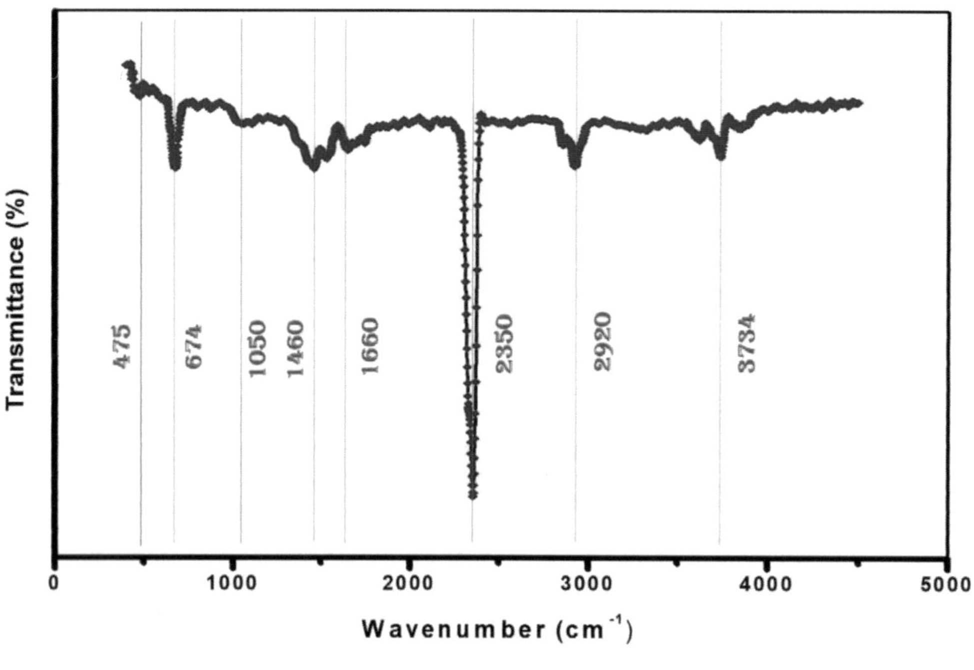

Figure 3. FTIR spectra of Ho coated ZnO nanorods.

3.4. UV-Visible spectroscopy

The optical absorption spectra of Holmium coated ZnO was explored by UV-Vis spectrometer in the wavelength range (280-400nm) shown in the Figure 4. The three absorption bands at 314, 356 and 382 nm were blue shifted as compared with bulk absorption band of ZnO; band gap emission of ZnO originates from the recombination of the free exciton [33]. The band located at 382 nm may be attributed to $\pi \rightarrow \pi^*$ transitions of ZnO; from the valence band to the conduction band ($O_{2p} \rightarrow Zn_{3d}$) as shown in the schematic diagram (Figure 5). The Two other bands located at 356 nm and 314 nm may be assigned to

4f→4f transitions (n→ π* and π → π* transitions) of Ho that are located at the surface of the ZnO nanorods, which give rise to the energy transfer from photon generated electron-hole pair in the ZnO nanorods to Ho in the surface of the ZnO nanorods [34]. The schematic diagram in figure 5 illustrates different transitions and absorption mechanisms indicated in the UV-Vis spectra. The blue shift may be ascribed to the nano size effect of the synthesized ZnO or may be attributed to the higher defect density on the surface of nanorods as shown in the SEM images, which may be expected to provide new bound excitons with new binding energies, which gave rise to the appearance of new absorption peaks [30]. Or maybe ascribed to the combination of optical transition to the excitonic state of ZnO nanorods and electronic transitions involving the 4f levels in Ho substituting the Zn ions [35].

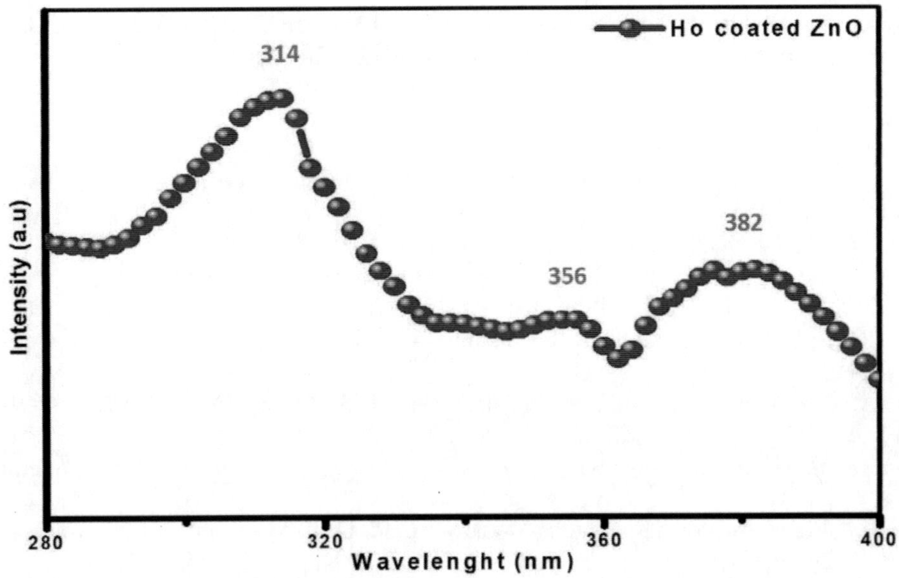

Figure 4. UV-Vis spectra of Ho coated ZnO nanorods.

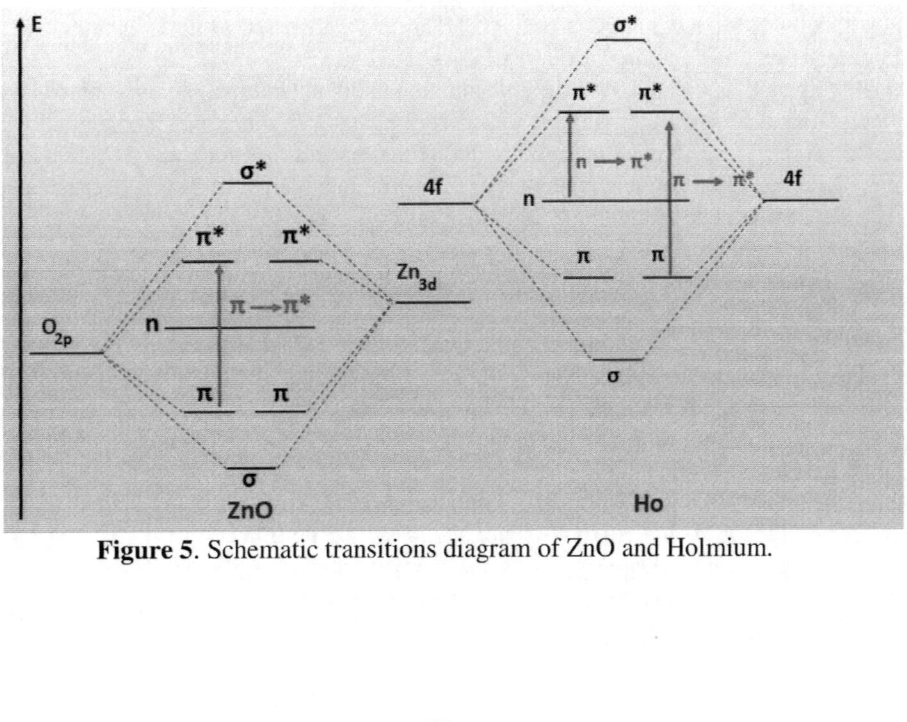

Figure 5. Schematic transitions diagram of ZnO and Holmium.

Conclusion
To summarize, to the best of our knowledge, the present study explores, for the first time, the synthesis and characterization of ZnO nanorods coated with Holmium. ZnO nanorods were prepared using hydrothermal method, the surface of ZnO nanorods was coated with Holmium by drop coating. Structural and morphological properties have been performed. The results revealed that both pure and Ho coated ZnO have the hexagonal wurtzite structure, nanorods were grown with different sizes and coated tightly with holmium. UV-Vis spectra measurements demonstrated that the coated material exhibits blue shifted absorptions peaks. This study was focused to reveal information about the structural and optical properties of Yb doped ZnO nanorods, which may enhance their performance for the optoelectronic applications.

Acknowledgments
This work was supported by the Faculty of Sciences Dhar el Mahraz USMBA (Morocco), University of South Africa Department of Physics (South Africa) and Africa Graphene Center (AGC).

References
[1] Klingshirn C 2007 *Chem. Phys. Chem.* **8** 782–803
[2] Ozgur U, Alivov Y I, Liu C, Teke A, Reshchikov M A, Dogan S, Avrutin V, Cho S J and Morkoç H 2005 *J. Appl. Phys.* **98** 041301
[3] Look D C 2001 *Materials Science and Engineering* **B80** 383–387
[4] Aydogu S, Sendil O and Coban M 2012 *Chinese Journal Of Physics* **50** 89-100
[5] Bai S, Liu X, Li D , Chen S, Luo R and Chen A 2011 *Sensors and Actuators* **B 153** 110–116
[6] Roso S, Güell F, Martínez-Alanis P R, Urakawa A, Llobet E 2016 *Sensors and Actuators* **B 230** 109–114
[7] Asthana A, Momeni K, Prasad A, Yap Y K and Yassar R S 2011 *Appl Phys A* **105** 909–914
[8] Madlol R A A 2017 *Results in Physics* **7** 1498–1503
[9] Edalati K, Shakiba A, Vahdati-Khaki J and Zebarjad S M 2016 *Materials Research Bulletin* **74** 374–379
[10] Laurenti M, Garino N, Porro S, Fontana M and Gerbaldi C 2015 *Journal of Alloys and Compounds* **640** 321-326
[11] Matinise N, Fuku X G, Kaviyarasu K, Mayedwa N and Maaza M 2017 *Applied Surface Science* **406** 339-347
[12] Foo K L, Hashim U, Muhammad K and Voon C H 2014 *Nanoscale Research Letters* **9** 429
[13] Alimanesh M, Hassan Z and Zainal N 2017 *Optical Materials* **72** 276-282
[14] Kim S J, Kim H H, Kwon J B, Lee J G, Hoan B O, Lee S G, Lee E H and Park S G 2010 *Microelectronic Engineering* **87** (5 – 8) 1534 – 1536
[15] Nur S R, Zainovia L, Azlan A A and Khairunisak A R 2016 *Materials Science Forum* **846** 459-464
[16] Lim S K, Hong S H, Hwang S H, Kim S and Park H 2013 *J. Mater. Sci. Technol.* **29** 39-43
[17] Kafle B P, Acharya S, Thapa S and Poudel S 2016 *Ceramics International* **42** 1133-1139
[18] Petersen J, Brimont C, Gallart M, Schmerber G, Gilliot P, Ulhaq-Bouillet C , Rehspringer J L, Colis S, Becker C, Slaoui A and Dinia A 2010 *J. Appl. Phys.* **107** 123522
[19] Babu S and Ratnakaram Y C 2016 *AIP Conference Proceedings* **1731** 070001
[20] Zamiri R, Lemosa A F, Rebloa A, Ahangarb H A and Ferreira J M F 2014 *Ceramics International* **40** 523–529
[21] Senol S D 2016 *J Mater Sci: Mater Electron* **27** 7767–7775
[22] Soumahoro I, Schmerber G, Douayar A, Colis S, Abd-Lefdil M, Hassanain N, Berrada A, Muller D, Slaoui A, Rinnert H and Dinia A 2011 *J. Appl. Phys.* **109** 033708.
[23] Khataee A, Saadi S, Vahid B and Joo S W 2016 *Journal of Industrial and Engineering Chemistry* **35** 167-176
[24] Singh S, Deepthi J N D, Ramachandran B and Rao M S R 2011 *Materials Letters* **65** 2930–2933
[25] Franco A and Pessoni H V S 2016 *Materials Letters* **180** 305-308
[26] Li X and Wang Y 2011 *Journal of Alloys and Compounds* **509** 5765–5768
[27] Chen B, Meng C, Yang Z, Li W, Lin S, Gu T, Guo X, Wang D, Yu S, Wong C W and Tong T 2014

Optics Express **22**(20) 24276-24285
[28] Panda N R, Sahu D, Acharya B S and Nayak P 2015 *Current Applied Physics* **15** 389-396
[29] Ameen S, Akhtar M S and Shin H S 2012 *Chemical Engineering Journal* **195–196** 307–313
[30] Soni B H, Deshpande M P, Bhatt S V, Garg N and Chaki S H 2013 *J. NANO- ELECTRON. PHYS.* **5** 04077
[31] Xiong G, Pal U, Serrano J G, Ucer K B and Williams R T 2006 *phys. stat. sol.* **3** No.10 3577–3581
[32] Pandey P, Kurchania R and Haque F Z 2014 *Journal of Advanced Physics* **3** 1–7
[33] He X, Qian H, Zhi Q, Zhang M, Luo J, He R and Zeng Q 2013 *Applied Surface Science* **283** 571-576
[34] Zeng X, Yuan J and Zhang L 2008 *J. Phys. Chem. C* **112** No 10 3503-3508
[35] Sharma P K, Dutta R K and Pandey A C 2010 *Journal of Colloid and Interface Science* **345** 149–153

Removal of hexavalent chromium from wastewater using PPy/Fe_3O_4 magnetic nanocomposite influenced by rotating magnetic field from two pole three-phase induction motor

U O Aigbe[1], W H Ho[2], A Maity[3], M Khenfouch[4], and V Srinivasu[1]

[1] Department of Physics, College of Science, Engineering and Technology, University of South Africa, Pretoria 0001, South Africa.
[2] Department of Industrial and Mechanical Engineering, College of Science, Engineering and Technology, University of South Africa, Pretoria 0001, South Africa.
[3]DST/CSIR National Centre for Nanostructured Materials, Council for Scientific and Industrial Research, Pretoria 0001, South Africa
[4]Africa Graphene Centre, Department of Physics, College of Science, Engineering and Technology, University of South Africa, Pretoria 0001, South Africa

E-mail: uyi4we@yahoo.co.uk

Abstract. The influence of varying rotating magnetic field using a 2-pole three-phase induction motor on the removal of hexavalent chromium ions from wastewater using polypyrrole magnetic nanocomposite was explored in this study. Hexavalent chromium removal in this study was observed to be pH dependent under the influence of rotating magnetic field, as the percentage removal of hexavalent chromium decreased with increase in pH. The percentage amount of hexavalent chromium ions removed from the aqueous solution increased as the rotating magnetic field intensity was increased from 8.96-12.15 mT in the anticlockwise direction and 10.10-13.38 mT in the clockwise direction with maximum removals of 73% and 81% observed.

1. Introduction

Water bodies contaminated with toxic materials like heavy metals are prevalent in emerging countries due to increased industrial activities [1]. The occurrence of high concentration of heavy metals like chromium, mercury, cadmium, nickel, cobalt, and lead in natural water supplies and manufacturing wastewater streams is a crucial health and environmental issue due to their high toxicity and bioaccumulation through food chain [2].

Hexavalent chromium is termed a group 'A' human carcinogen owing to its mutagenic and carcinogenic properties. Its concentration in industrial wastewater ranges from 0.5-270,000 mg/L, with the tolerance limit for hexavalent chromium released into the inland surface water and potable water are 0.1 and 0.05 mg/L [3-4]. For the different existing treatment technologies for the removal of toxic heavy metals, adsorption, chemical precipitation, ion exchange, coagulation, reverse osmosis, electrolysis and membrane process are extensively used [5].

To reduce environmental problems resulting from heavy metals contamination, it is necessary to create new residue treatment methods, such that the residue recovery occurs in a faster and efficient way [6]. Magnetic field exposed method have progressively drawn attention to improving the adsorption and separation. Magnetism is an exclusive property that supports water purification as it influences the physical and chemical properties of the contaminants in an aqueous solution. [1, 7-8].

Magnetic field exposure method has shown its capacity for enhancing the removal of contaminants from water as static magnetization is appropriate, simple and cost-effective. Basically, the behaviour and physicochemical properties of water are affected by a magnetic field. With the application of magnetic field to water, the magnetic field within changes the rates of chemical reactions due to the existence of competing recations of dissolution and precipitation of dissolved salts facilitates the formation and decomposition of colloidal complexes and improves electro-coagulation followed by sedimentation and crystallization of scaling salts of Ca^{2+}, Mg^{2+}, Fe^{2+} and Fe^{3+}[9-10].

Increase in the application of an electromagnetic field in different fields including therapeutic and diagnostic medicine, environment management and industrial procedures have been witnessed in recent years [11]. The magnetic field was reported to have significant effects on scale prevention, [12-14], suspended particle separation [15], adsorption of zinc, copper (II), industrial waste, colour removal, total suspended solids, uranium and strontium [16-21].

The motivation for this study is to evaluate the effect of magnetic field on polypyrrole magnetic nanocomposite for the adsorption of Cr(VI) ion. The goal is to develop a water treatment procedure with the information obtained, which would be useful for further application in the treatment of industrial effluents.

2. Materials and method

2.1. Materials

Magnetite (Fe_3O_4), anhydrous ferric chloride ($FeCl_3$), pyrrole (Py), potassium dichromate ($K_2Cr_2O_7$), sodium hydroxide (NaOH), hydrochloric acid (HCI) used for all experiments were of analytical grade and purchased from Sigma-Aldrich, South Africa. A stock solution of Cr(VI) was prepared by dissolving 2.83 g of $K_2Cr_2O_7$ in deionized water. For further experiments, required solutions of Cr(VI) were made by diluting the appropriate stock solution with deionized water and altering the solution pH using 0.1 M HCI or NaOH.

2.2. Synthesis of Polypyrrole Magnetic Nanocomposite

Polypyrrole magnetic nanocomposite (PPy/Fe_3O_4) was synthesized via the in-situ chemical oxidative polymerization procedure as described by [2]. 0.4 g of magnetite (Fe_3O_4) was added to 80 mL deionized water in a 250mL Erlenmeyer flask. To have a good distribution of Fe_3O_4 in the deionized water, it was ultrasonicated for 30 minutes. 0.8 mL of pyrrole monomer was then syringed into the mixture after ultrasonication, with 6 g of oxidant ($FeCl_3$) added afterwards. The mixture was shaken manually for 5 minutes to make a homogeneous mixture and kept for 3 hours for complete polymerization to take place at room temperature. The black powder obtained was filtered and washed with deionized water until the filtrate becomes colourless. To stop the reaction of the mixture, acetone was added. The synthesized nanocomposite was dried for 24 hours under vacuum at 60 °C.

2.3. Characterization of polypyrrole magnetic nanocomposite (PPy/Fe_3O_4)

The size, morphology and elementary composition of polypyrrole magnetic nanocomposite were acquired from a JEOL JEM-2100 high-resolution transmission electron microscopy with an accelerating voltage of 200 kV and Leo-Zeiss scanning electron microscopy equipped with energy dispersive X-ray (EDX). The surface area analysis was measured using Brunauer Emmett-Teller (BET) at low-temperature nitrogen adsorption-desorption technique on a Micromeritics ASAP 2020 gas adsorption apparatus (Micromeritics USA). The X-Ray Diffraction (XRD) patterns of the PPy/Fe_3O_4 nanocomposite were studied using SmartLab X-ray diffractometer operating at 45 kV/200 mA with Cu-

Kα radiation (wavelength λ=1.540593 Å) in the 2θ range of 10-70°. FTIR spectra were measured using a Perkin-Elmer Spectrum 100 spectrophotometer with a scanning range of 500-3000 cm^{-1}. The magnetic property of the nanocomposites before and after adsorption of Cr(VI) was confirmed by Bruker-Electron Spin Resonance (ESR) spectrometer at room temperature. The vibrating sample magnetometer of the physical property measurement system (PPMS) by Quantum Design temperature was applied to determine the magnetic properties of the nanocomposite at an interval of -10 000 to 10 000 Oe at room temperature.

2.4. Experimental setup

The experimental setup comprises of a three-phase induction motor, a power source, a toroidal cleaning chamber, a thermocouple and a fan. The nameplate parameters of the three-phase induction are shown in table 1. The three-phase induction motor has a stator winding with diameter and height of 15 x 160 mm. The rotor is modified to hold an open-top toroidal cleaning chamber made from perspex plastic. The open top toroidal cleaning chamber used for the adsorption process influenced by magnetic field had an internal, external diameters and height of 66 x 100 x 25 mm respectively. To supply power to the stator windings of the three-phase induction motor, a 10 A variable power supply was used. The magnetic field created in the air-gap of the induction motor was adjustable with the input power supplied to the stator windings, with a maximum magnetic field of 13.2 mT generated. The magnetic field created was measured by a means of digital GM07/GM08 gaussmeter. To dissipate heat generated from the stator windings onto the system, a 220 V, 34 W fan mounted on a stand and held above the three-phase induction motor was used. A graphic diagram of magnetic field adsorption device is shown in figure 1.

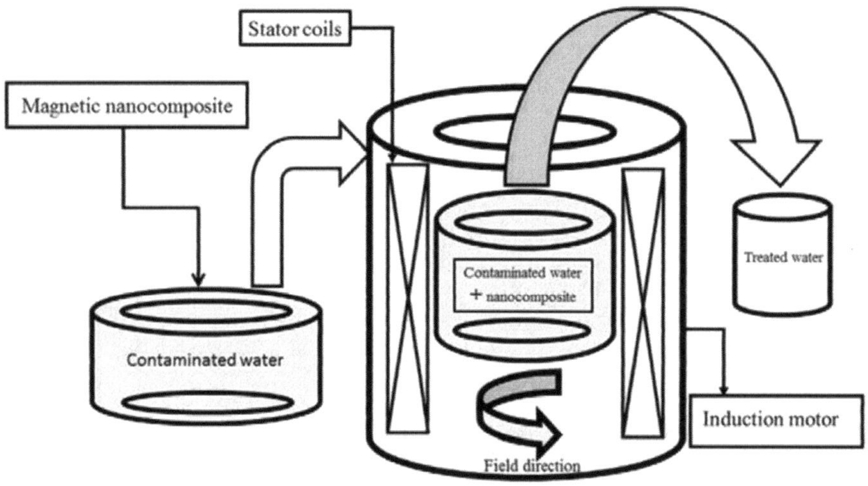

Figure 1. Schematic of a two-pole magnetic field adsorption device.

Table 1. 2-Pole three-phase induction motor specification.

Specification	Value/Unit
Horsepower	1.5 kW
Frequency	50 Hz
Synchronous speed	2920 RPM
Voltage	380/400 V
Current	2.19 A

2.5. Adsorption studies

Adsorption experiments were performed at room temperature in a 2-pole three-phase induction motor (first electric induction motor) under magnetic field intensity range of 9.0-13.2 mT. For pH effect on the adsorption of Cr(VI) under the influence of magnetic field, pH of Cr(VI) solution was varied from 2-10, which was spun under a constant magnetic field of 13.38 mT. For adsorbent dosage effect on Cr(VI) adsorption under the influence of a magnetic field, adsorbent dosage was varied from 0.01-0.05 g at a constant pH and magnetic field of 13.38 mT. Magnetic field effect on the adsorption of the Cr(VI) experiments was carried out by varying the magnetic field in clockwise and anticlockwise directions. At the end of all experiments, samples are taken out with a syringe and filtered through a 0.45 µm cellulose acetate syringe filter. The residual concentration of Cr(VI) ions in the filtrate were analyzed using a UV/Vis spectrophotometer (PerkinElmer Lambda 1050) at a wavelength of 540 nm by complexing the Cr(VI) with 1,5 diphenylcarbazide regent. Each experiment was conducted in triplicate and the average values reported. The percentage of Cr (VI) removal (adsorption efficiency) was calculated according to equation 1:

$$Cr^{6+}\ \%\ \text{Removal} = \left(\frac{C_0 - C_e}{C_0}\right) \times 100 \qquad (1)$$

where C_0 and C_e is the initial and equilibrium concentration of Cr(VI) concentration in mg/L.

3. Results and discussion

3.1. Characterization of PPy/Fe$_3$O$_4$

The size and surface morphology of PPy/Fe_3O_4 nanocomposite were observed using HR-TEM and SEM (figure 2). The HR-TEM image shown in figure 2 (a) confirms that the nanocomposites had a regular uniform spherical morphology with average diameters of 10.3$\pm$ 8.3 nm. The surface analysis results (figure 2 (b)), shows that the Fe_3O_4 nanoparticles are fully covered and stabilised by the PPy microspheres in the PPy/Fe_3O_4 composite matrix. In addition, the energy-dispersive x-ray analysis (EDX) spectrum of PPy/Fe_3O_4 confirms the presence of C, O, N, CI and Fe as the major elements in the polypyrrole magnetic nanocomposite (figure 2 (c)). After Cr(VI) adsorption, an additional Cr peak along with C, O, N, CI and Fe is observed. This confirms the adsorption of Cr(VI) ion onto PPy/Fe_3O_4 nanocomposite surface (figure 2 (d)).

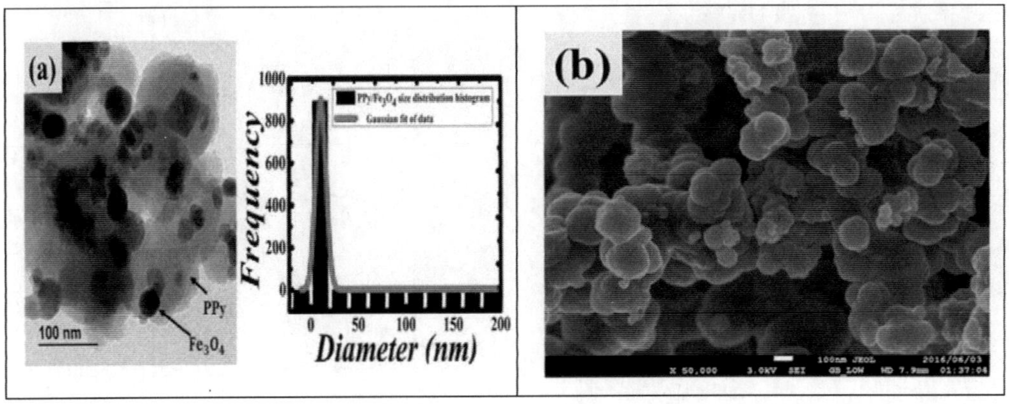

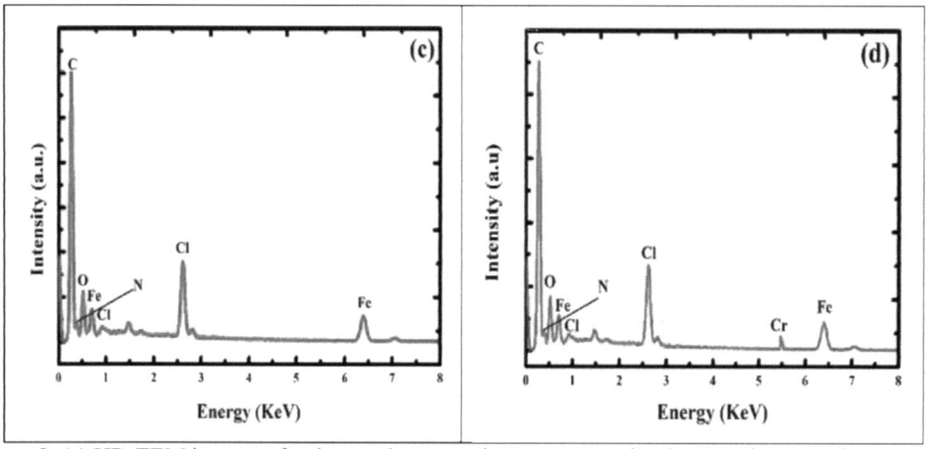

Figure 2. (a) HR-TEM images of polypyrrole magnetic nanocomposite (b) SEM images of polypyrrole magnetic nanocomposite (c) EDX spectrum of PPy/Fe_3O_4 nanocomposite before hexavalent chromium adsorption and (d) EDX spectrum of PPy/Fe_3O_4 magnetic nanocomposite after hexavalent chromium adsorption.

The nitrogen adsorption-desorption isotherm for the nanocomposite is shown in figure 3. The result reveals hysteresis loop that closes near a relative pressure of 0.82. This is indicative of mesoporous characteristics of the nanocomposite. The adsorption-desorption isotherm shape indicates a typical type-IV, which agrees with the nature of mesoporous adsorbents according to IUPAC [22]. The specific surface area of the nanocomposite was 28.77 m^2/g, while the BJH cumulative volume of pores and BJH average pore diameter were determined to be 0.06 cm^3/g and 15.82 nm, respectively.

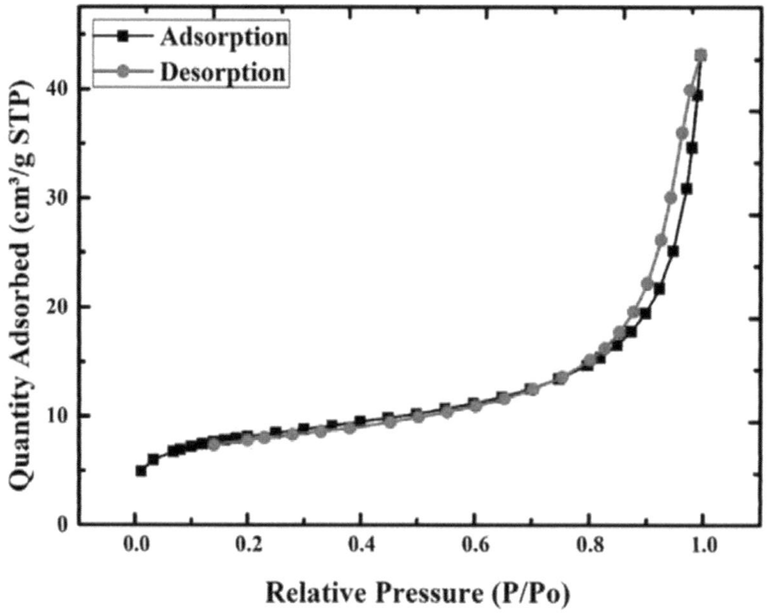

Figure 3. N_2 adsorption-desorption isotherms for PPy/Fe_3O_4 nanocomposite.

The XRD diffraction patterns of PPy, Fe_3O_4 nanoparticles, and the polypyrrole magnetic nanocomposite before and after adsorption of Cr(VI) is shown in figure 3. A broad peak (figure 4 (a)) observed at $2\theta =$

24.6°, is a characteristic peak of amorphous PPy [23]. The characteristic diffraction peaks of Fe_3O_4 observed at $2\theta = 30.3°, 35.7°, 43.4°, 53.6°, 57.3°$, and $62.9°$ are assigned to (220), (311), (400), (422), (511), and (440) planes of Fe_3O_4 [24-27]. Peaks of Fe_3O_4 were observed in the nanocomposites before and after adsorption of Cr(VI) (figure 4 (c)). This shows that after modification of Fe_3O_4 nanoparticles, their phases were not modified. The XRD study clearly shows that the Fe_3O_4 nanoparticles were successfully incorporated into the polypyrrole network.

The XRD diffraction patterns of PPy, Fe_3O_4 nanoparticles, and the polypyrrole magnetic nanocomposite before and after adsorption of Cr(VI) is shown in figure 3. A broad peak (figure 4 (a)) observed at $2\theta =$ 24.6°, is a characteristic peak of amorphous PPy [23]. The characteristic diffraction peaks of Fe_3O_4 observed at $2\theta = 30.3°, 35.7°, 43.4°, 53.6°, 57.3°$, and $62.9°$ are assigned to (220), (311), (400), (422), (511), and (440) planes of Fe_3O_4 [24-27]. Peaks of Fe_3O_4 were observed in the nanocomposites before and after adsorption of Cr(VI) (figure 4 (c)). This shows that after modification of Fe_3O_4 nanoparticles, their phases were not modified. The XRD study clearly shows that the Fe_3O_4 nanoparticles were successfully incorporated into the polypyrrole network.

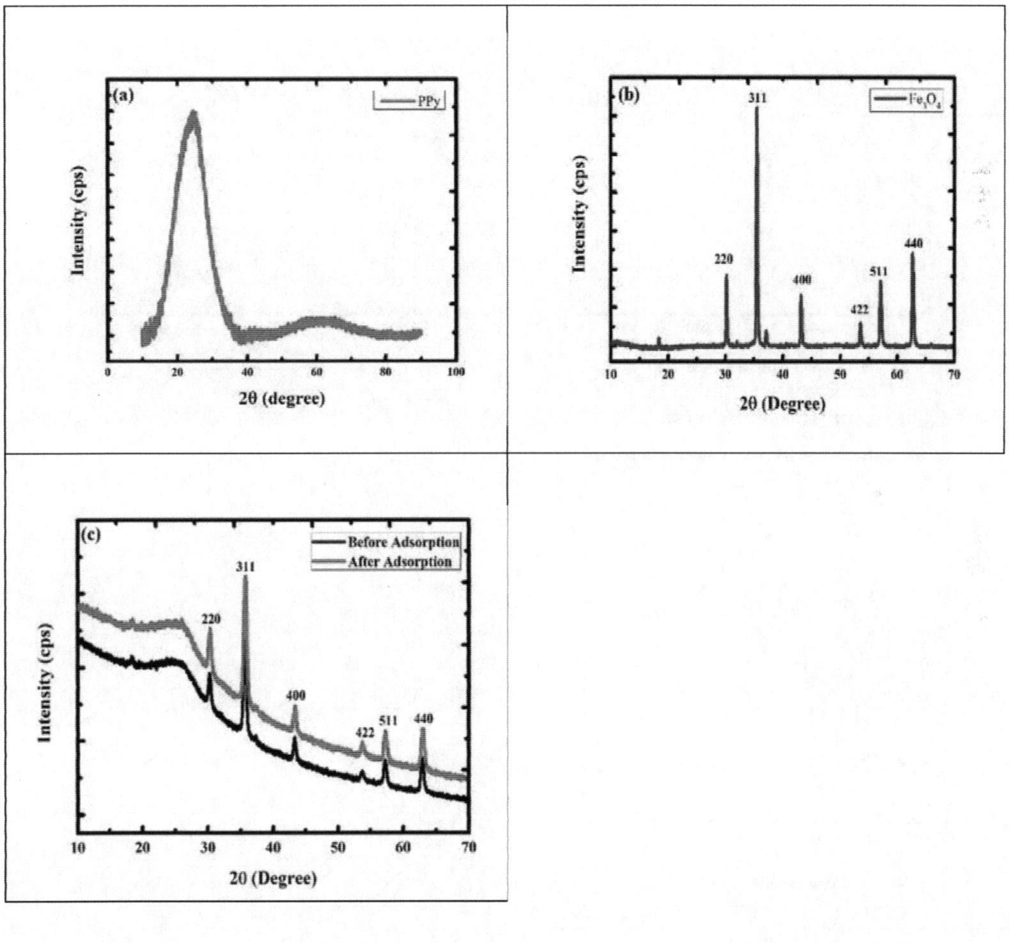

Figure 4. XRD pattern of (a) polypyrrole, (b) magnetite (Fe_3O_4), and (c) polypyrrole magnetic nanocomposite before and after adsorption of hexavalent chromium using 2-pole three-phase induction motor.

The FTIR spectra of Fe_3O_4, PPy, polypyrrole magnetic nanocomposite before and after Cr(VI) adsorption obtained from 500-2800 cm^{-1} is shown in figure 5. The characteristic peak of Fe_3O_4

nanoparticles appears at 549 cm^{-1} (figure 5 (b)), which is due to the stretching vibration mode associated to the metal-oxygen adsorption band (Fe-O bonds in the crystalline lattice of Fe_3O_4) [28]. The bands observed at 1534 and 1449 cm^{-1} relate to the fundamental vibration of the pyrrole rings [29-30]. The bands at 1040 and 1157 cm^{-1} corresponds to =C-H out of plane vibration [31]. Peaks observed at 967-853 cm^{-1} are attributed to C-H deformation [2]. After adsorption of chromium (VI) ions, there were clear shifts in the peaks values to the higher infrared region. The FTIR spectrum of PPy/Fe_3O_4 nanocomposite after adsorption peaks were observed at 1537, 1460, 1159, 1042, 971 and 895 cm^{-1}, which were attributed to the absorption bands of symmetric and antisymmetric ring-stretching, conjugated C-N stretching, C-H deformation and C-H stretching vibration of pyrrole ring, respectively. The shift in the adsorption peaks (after adsorption) indicated that there was a metal binding process taking place at the adsorbent surface [2, 32].

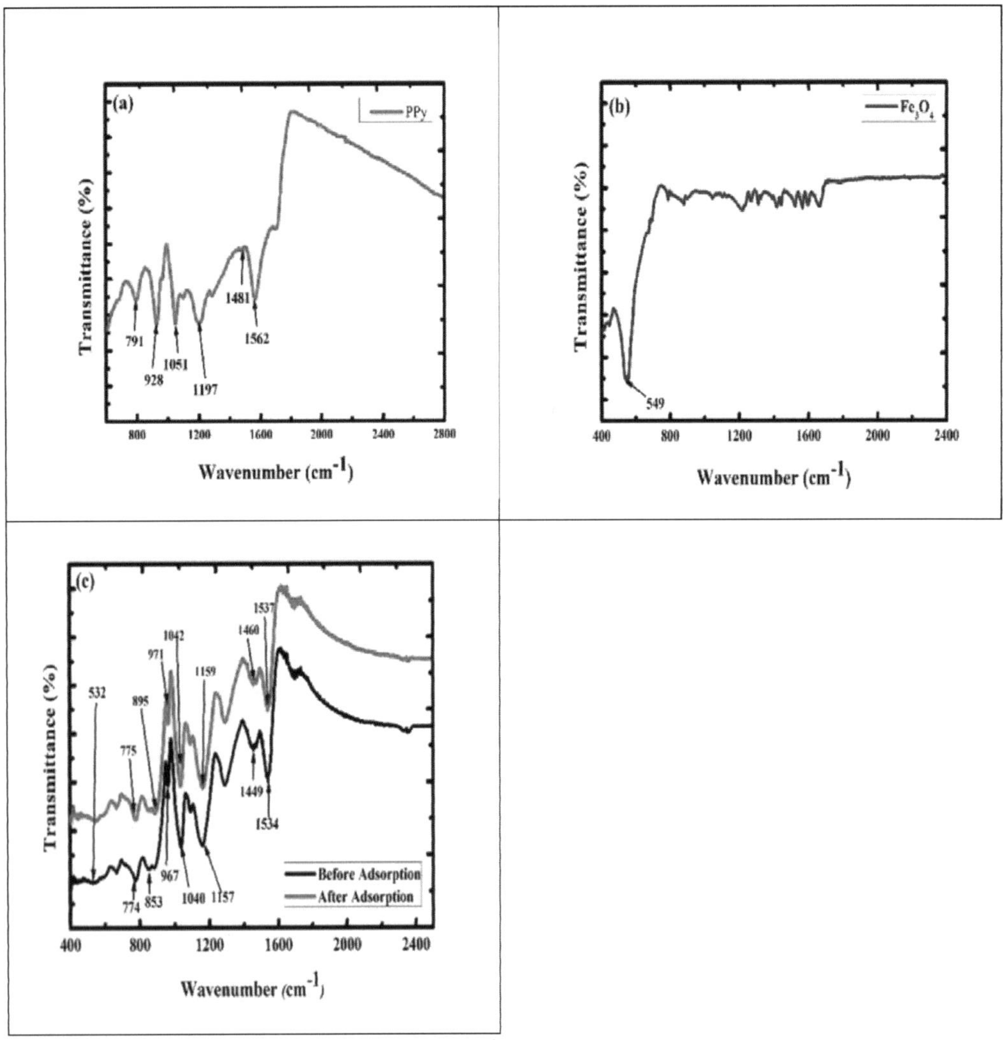

Figure 5. FTIR spectra of (a) PPy, (b) Fe_3O_4 and (c) polypyrrole magnetic nanocomposite before and after adsorption of hexavalent chromium using 2-pole three phase induction motor.

The ESR adsorption spectra for polypyrrole magnetic nanocomposite before and after adsorption of Cr(VI) at room temperature is shown in figure 6. The line shapes were observed to be symmetrical. The resonance signal was also observed to be wide and broad with a linewidth value of 1400 Gauss. The effective g-value of the nanocomposite was determined to be 2.25 G, which was attributed to Fe^{3+} spin interactions. Such interactions show a superparamagnetic behaviour characterized by the presence of clusters. The shapes and the field position are identical to those of the standard magnetic nanoparticles suspension and are consistent with ESR spectra of superparamagnetic iron oxide nanoparticles at room temperature [33-35]. The line shape of polypyrrole magnetic nanocomposite when fitted to Gaussian and Lorentzian functions, showed a line shape having a combination of 55% Gaussian and 45% Lorentzian functions. The dominant Gaussian fit is characteristic of ferromagnetic resonance indicating that the main percentage of the polypyrrole magnetic nanocomposite is magnetic as shown in figure 6 (b), as this confirms that the dominated fraction of the PPy/Fe_3O_4 is magnetic [36-37].

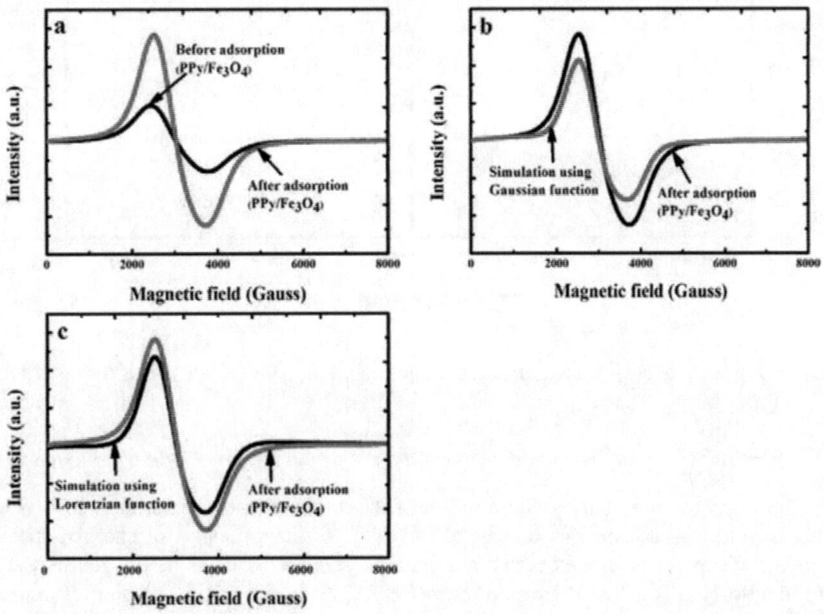

Figure 6. ESR spectra of PPy/Fe_3O_4 (a) before and after adsorption of hexavalent chromium, (b) PPy/Fe_3O_4 nanocomposite after Cr(VI) adsorption **(b)** ESR spectra of PPy/Fe_3O_4 nanocomposite after hexavalent chromium adsorption fitted with Gaussian function and **(c)** ESR spectra of PPy/Fe_3O_4 nanocomposite after hexavalent chromium adsorption fitted with Lorentzian function.

The magnetization hysteresis loop of polypyrrole magnetic nanocomposite at room temperature (300 K) was measured at intervals of -10 000 to 10 000 is shown in figure 7. The saturation magnetization was determined to be 23 emu/g, with a zero-remanence magnetization and coercivity field, which is typical of a superparamagnetic material. The superparamagnetic behaviour is mostly generated by the magnetite core in the sample [38].

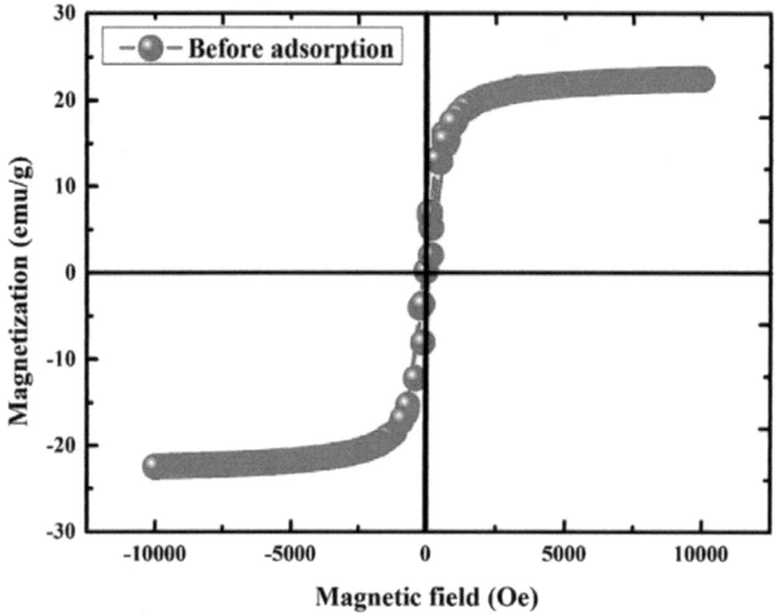

Figure 7. Magnetization curve of polypyrrole magnetic nanocomposite at room temperature.

3.2. Effect of pH on adsorption of hexavalent chromium under the influence of magnetic field

Determining the point of zero charge where the net surface charge of a material is zero is vital for understanding the electrostatic interactions at the materials surface, mostly for charged species. The adsorption of the adsorbate on non-specific and specific adsorbents depends on pH, as the surface property of the adsorbate is affected by solution pH [22, 39]. In adsorption medium, the pH of any solution is an important factor which controlled the metal ions adsorption onto the adsorbent surface [40-41].

The surface charge and the protonation degree of the adsorbent are significantly influenced by the pH value [42]. For electrostatic adsorption the zeta potential and its devolution with changing pH are important. For adsorption process, the pH is adjusted in a way that both components have opposed surface charges. By changing the pH-value and hence the zeta potential of one component repulsive interaction forces occurs and the molecules detach from the surface of the carrier particles. One of the most important parameters that influence adsorption of metals ions is the solution pH, which influences the surface properties the adsorbents and ionic forms of the chromium solutions [43].

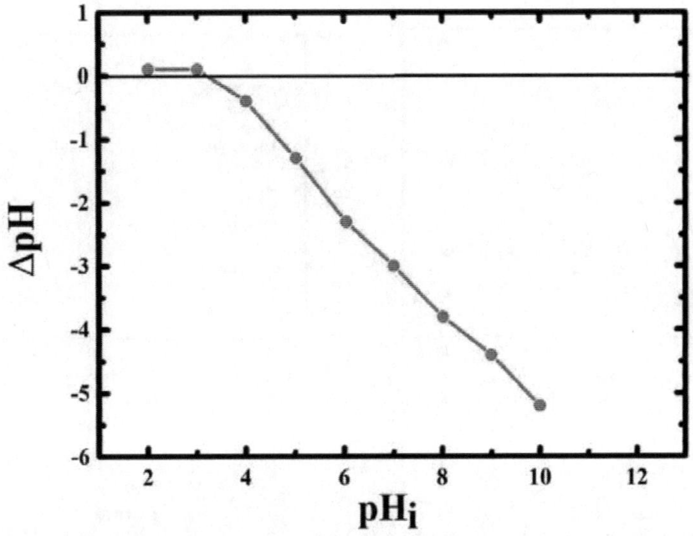

Figure 8. Point of zero charge of polypyrrole magnetic nanocomposite.

Increase in the percentage of Cr(VI) removed as the pH is increased from 2-10 is observed, with a maximum removal observed at pH 2 is shown in figure 9 (a). It is remarkable that Cr(VI) removal curve can be explained by the surface properties of the polypyrrole magnetic nanocomposite and the relative distribution Cr(VI) species under a wide range of pH value [44]. The pH_{pzc} of the nanocomposite was found to be 3.20 as shown in figure 8. The surface of the magnetic nanocomposite was mainly negative charged when the pH > pHpzc (3.20) and mainly positive charged when pH < pHpzc (3.20). When the solution pH was below the pH_{pzc} (< 3.20), Cr(VI) ions were electrostatically attracted to the positively charged magnetic nanocomposite, as the Cr(VI) exists as $HCrO_4^-$ in anionic species at acidic pH, hence the enhanced Cr(VI) removal at the acidic condition. With increase in solution pH, low percentage of Cr(VI) removed was due to the electrostatic repulsion between the anionic species of Cr(VI) (CrO_4^{2-}) and negative charged OH^- ions on the surface of the polypyrrole magnetic nanocomposite, causing a decrease in the efficiency of the adsorbent for Cr(VI) ions removal.

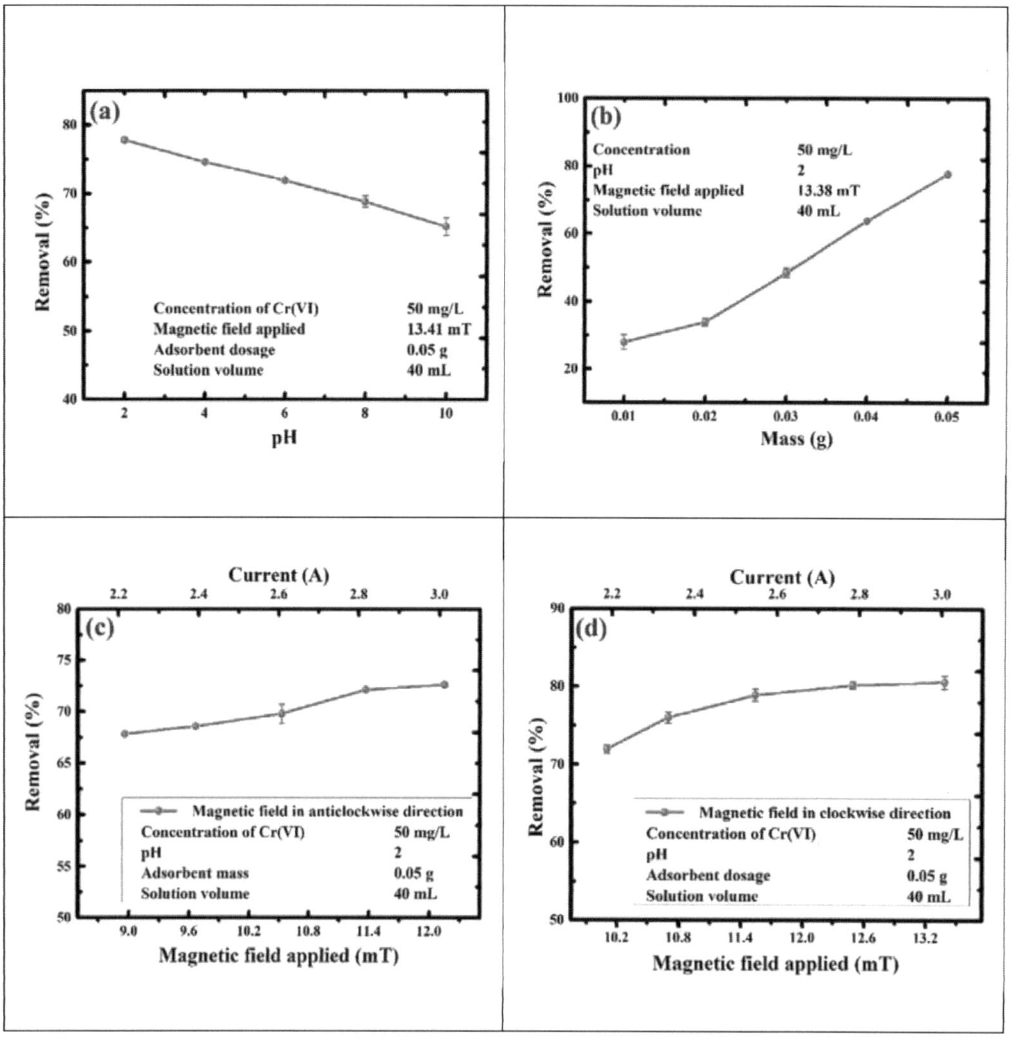

Figure 9. Effect of (a) pH on the adsorption of hexavalent chromium onto PPy/Fe_3O_4 nanocomposite, (b) adsorbent dosage on the adsorption of hexavalent chromium onto PPy/Fe_3O_4 nanocomposite, (c) magnetic field strength on the adsorption of hexavalent chromium onto PPy/Fe_3O_4 nanocomposite (anticlockwise direction) and (d) magnetic field strength on the adsorption of hexavalent chromium onto PPy/Fe_3O_4 nanocomposite (clockwise direction).

3.3. Effect of adsorbent dosage on adsorption of hexavalent chromium under the influence of magnetic field

The effect of adsorbent dosage on the removal of hexavalent chromium by the magnetic nanocomposite was investigated under the influence rotating magnetic field from a two-pole induction motor at room temperature. The results indicated in figure 9 (b) shows that the percentage of Cr(VI) removed using the magnetic nanocomposite under the influence of rotating magnetic field increased with increase in the nanocomposite dosage. Precisely, the range of Cr(VI) removed from 40 mL of 50 mg/L Cr(VI) solution increased from 27.9% at a dosage of 0.01 g to 77.8% at a dosage of 0.05 g. Increase in the amount of hexavalent chromium removed from aqueous solution with an increase in the adsorbent dosage was due to increase in the number of active adsorption sites for a constant concentration of hexavalent chromium. At low adsorbent dosage, the active adsorption sites were insufficient for the adsorbate (Cr(VI)) to occupy. However, at high adsorbent dosage, the active adsorption sites are sufficient for the adsorbate to occupy. Similar results have been reported in the study of Cr(VI) adsorption [43, 45].

3.4. Effect of magnetic field on adsorption of hexavalent chromium

The effect of the induced magnetic field in anticlockwise and clockwise direction on the adsorption of Cr(VI) ions onto polypyrrole magnetic nanocomposite are shown in figure 9 (c and d). It is observed that percentage of Cr(VI) removed onto polypyrrole magnetic nanocomposite under the influence of rotating magnetic field increased from 72-81% and 68-73% with the increase in magnetic field range of 8.96-13.38 mT in the clockwise and anti-clockwise directions. Improved removal of Cr(VI) ions under magnetic field were attributed to the increased size of particles aggregates as the induced magnetic field on the polypyrrole magnetic nanocomposite was increased from low magnetic field to high magnetic field. The sizes of aggregates formed were observed to depend on the magnetic velocity, with fast moving uniform particles aggregates with a large surface area for adsorption process were observed at high magnetic field. Smaller non-uniform particles aggregates travelling with slow drift velocities were also observed at the low magnetic field, with a reduced surface area for the adsorption process. Chain breakage in the aggregates was also noticed to significantly reduces the swimming efficiency of the particles at low magnetic field, as the segment of the chain broken were incapable to rejoin the original chain formation, leading to the reduced surface area for the adsorption of Cr(VI) ions. The lateral binding of chains driven into contact by magnetic manipulations substantially increases the chain polydispersity, hence increased chain collision and area of particle interaction [46-47].

5. Conclusion

The effect of rotating magnetic field on the adsorption of Cr(VI) onto polypyrrole magnetic nanocomposite was explored in this study. The study showed that Cr(VI) adsorption onto the polypyrrole magnetic nanocomposite under the influence of magnetic field was dependent on the solution pH, adsorbent dosage and the magnetic field direction. Enhanced percentage of Cr(VI) ions removed was observed at pH 2, at an adsorbent dosage of 0.05 g and magnetic field of 12.15 mT (anticlockwise direction) and 13.38 mT (clockwise direction). Increased in the magnetic field was also observed to enhance Cr(VI) ions removal. These were due to increased aggregation of particles. The size of aggregates formed depended on the magnetic velocity, with fast moving large uniform particles aggregates observed. Smaller non-uniform particles aggregates travelling with slow drift velocities were also observed at the low magnetic field.

Acknowledgements

The authors are thankful for the financial support from the Research and Innovation Directorate of UNISA through the Innovation Support Programme, National Centre for Nanostructured Materials, South Africa Council for Scientific and Industrial Research, Pretoria, Dr M. Bhaumik and Dr RAGHUNATH DAS for their support and advice.

References

[1] Vazquez O F G, Virgen M M, Montoya V H, Gomez R T, Flores J L A, Cruz M P and Moran M A M 2016 *Indust. and Engr. Chem. Res* **55** 9323-9331.
[2] Bhaumik M, Maity A, Srinivasu V V, and Onyango M S 2011 *J. of Hazard. Materials* **190** 381-390.
[3] Attia A A, Khedr S A and Elkholy S A 201, *Brazilian J. of Chem. Engr.* **27** 183-193.
[4] Nethaji S, Sivasamy A and Mandal A B 2013 *Bioresource. Tech.* **134** 94-100.
[5] Ilankoon N 2014 *J. of Engr Res. and Appln.* **4**55-63.
[6] De Brito J F, De Oliveira Ferreira L, Pereira M C R, Da Silva J P and Ramalho T C 2012 *Water, Air, and Soil Polltn.* **223** 3545-3551.
[7] Hao X, Liu H, Zhang G, Zou H, Zhang Y, Zhou M and Gu Y 2012 *Applied Clay Sci.* **55** 177-180.
[8] Duan L, Guo S and Yang J 2012 *Adv. in Chem. Engr and Sci* **2** 101-107.
[9] Li G, Zhu W, Zhang C, Zhang S, Liu L, Zhu L and Zhao W 2016 *Bioresource. Tech* **206** 16-22.

[10] Mosin O and Ignatov I 2014 *European J. of Mole. Biotech* **2** 72-85.
[11] Ali Y, Samaneh R, Zohre R and Mostafa J 2014 *Curr. World Environ* **9** 1008-1016.
[12] Quinn C J, Molden T C and Sanderson C H 1997 *Iron and Steel Engnr.* **74** 47-52.
[13] Holysz L, Szczes A and Chibowski E 2007 *J. of Colloid. and Interface Sci.* **316** 996-1002.
[14] Higashitani K, Kage A, Katamura S, Imai K and Hatade S 1993 *J. of Colloid. and Interface Sci.* **156** 90-95.
[15] Khater Z Z K and Ibraheim M H 2015 *Intern. J. of Curr. Research and Academic Review* **3** 262-279.
[16] Zhang G, Yang X, Liu Y, Jia Y, Yu G, and Ouyang S, 2004, *J. of Colloid and Interface Sci.* **278** 265-269.
[17] Zhang G, Liu Y, Xie Y, Yang X, Hu B, Ouyang S, Liu H and Wang H 2005 *Applied Clay Sci.* **29** 15-21.
[18] Yuan-Yuan J, Gao-Ke Z, Bo H and Wei D 2004 *J. of Wuhan Uni. of Techno-Mater. Sc. Ed.* **19** 52-54.
[19] Mohammed R R, Ketabchi M R and Mckay G 2014 *Chem. Engr. J.* **243** 31-42.
[20] Rao M S and Sahu O 2013 *Phys and Matera. Chem.* **1** 34-40.
[21] Fondeur F F, Hobbs D T and Fink S D 2010 *Separation. Sci. and Techno.* **45**. 1876-1879.
[22] Mohamed M H, Pirlot M, Danquah M K and Wilson L D 2017 *J. of Materials Sci. and Chem. Engr.* **5** 12-24.
[23] Azadmanjiri J, Hojati Talemi P, Simon G P, Suzuki K and Selomulya C 2011 *Polyr. Engr and Sci.* **51** 247-253.
[24] Shi D, Yang H, Ji S, Jiang S, Liu X and Zhang D 2015 *Procedia Engr.* **102**, 1555-1562.
[25] Silva V A J, Andrade P L, Silva M P C, Valladares L D L S and Aguiar J A 2013 *J. of Magntsm. and Magnetic Matera.* **343** 138-143.
[26] Mathad J K and Rao R M V G K 2011 *Polymer. Compts.* **32** 1416-1422.
[27] Peng W, Hu X and Zhang D 2011 *J. of Magnetism and Magnetic Maters* **323** 2064-2069.
[28] Qiu G, Wang Q and Nie M 2006 *Macromole. Maters and Engr.* **291**, 68-74.
[29] Li S, Lu X, Xue Y, Lei J, Zheng T and Wang C 2012 *PLoS One* **7** e43328.
[30] Tanzifi M and Karimipour K 2017 *J. of Water and Envirn. Nanotech.* **2** 265-277.
[31] Chen W, Li X, Xue G, Wang Z and Zou W 2003 *Appl. Surface Sci.* **218** 215-221.
[32] Sanches E A, Alves S F, Soares J C, Silva A M D., Silva C G D, Souza S M D, and Frota H O D 2015 *J. of Nanomaters.* **16** 301.
[33] Chertok B, Cole A J, David A E and Yang V C 2010 *Mol. Pharm.* **7** 375-385.
[34] Koseoglu Y, Yildiz F, Kim D K, Muhammed M and Aktas B 2004 *Phys. Stat. Sol.* **1** 3511-3515.
[35] Mamani J B, Gamarra L F and Brito G E D S 2014 *Mat. Res.* **17** 542-549.
[36] Bhaumik M, Leswifi T Y, Maity A, Srinivasu V V and Onyango M A 2011 *J. of Hazard. Maters* **186** 150-159.
[37] Kuznetsov O A, Sorokina O N, Leontiev V GShlyakhtin O A, Kovarski A L and Kuznetsov A A 2007 *J. of Magnetism and Magnetic Maters.* **311** 204-207.
[38] Silva V A J, Andrade P L, Silva M P C, Valladares, L D L S and Aguiar J A 2013 *J. of Magntsm. and Magnetic Maters.* **343** 138-143.
[39] Zou Y, Wang X, Ai Y, Liu Y, Li J, Ji Y and Wang X 2016 *Experimental Sci. and Techno.* **50** 3658-3667.
[40] Dui J, Zhu G and Zhou S 2013 *ACS Appl. Mater. Interfaces* **5** 10081-10089.
[41] Rao R A K, Khan A A and Alam N 2016 *Anal. Methods.* **8** 869-879.
[42] Anbia M, Kargosha K and Khoshbooei S 2015 *Chem. Engr. Research and Design* **93** 779-788.
[43] Mekonnen E, Yitbarek M and Soreta T R 2015 *S. Afr. J. Chem* **68** 45-52,
[44] Zhu K, Chen C, Xu H, Gao Y, Tan X, Alsaedi A and Hayat T 2017 *ACS Sustainable Chem. and Engr* **5** 6795-6802.
[45] Zhang R, Ma H, and Wang B, 2010, *Ind. Eng. Chem. Res.* **49**, 9998-10004.
[46] Aigbe U O, Das R, Ho W H, Srinivasu V and Maity A 2017, *Separtn. and Purification Techn.***194** 377-387.
[47] Kim D, Kim J and Choi W 2011 *J. of Hazardous Maters* **192** 928-931,

The decolourisation of Methyl Orange and textile effluent under UV using commercial and synthesized nano-TiO_2

N N Nyangiwe[1], B Baatjie[2], C Greyling[2], M Khenfouch[3], M Maaza[4,5]

[1] Natural Resources and the Environment, Source directed scientific measures group, CSIR, P O BOX 395, Pretoria, 0001
[2]Technology Station in Clothing and Textiles, Cape Peninsula University of Technology, Bellville
[3]Africa Graphene Center, Physics department, Eureka Building, College of Science, Engineering and Technology, University of South Africa, Florida, Roodepoort 1710.Gauteng
[4]UNISA Africa Chair in Nanosciences-Nanotechnology, College of Graduate Studies, University of South Africa, Muckleneuk ridge, PO Box 392, Pretoria, South Africa
[5] NanoAfNet, Nanolaboratories, iThemba LABS-National Research Foundation of South Africa, Old Faure Road, P O Box 722, Somerset West 7129, Western Cape Province, South Africa
Corresponding authors: khenfouch@yahoo.fr/nyangiwenangamso9@gmail.com

Abstract. The efficiency of the nanophotocatalysts was determined in a laboratory prepared solutions of model dye compound methyl orange (MO), which is a dyestuff typically applied on textiles. The reduction in the concentration of the dye over time was measured by using UV-VIS spectrophotometry. The decolourisation of MO solutions was characterized at the onset and after exposure to 9W germicidal UV lamps at various time intervals, viz. 30, 60, 120, 180, 300 and 420 minutes. The color transformation from dark-orange, (MO) to a transparent clear solution or a slightly colored solution is an indication of the decomposition of the organic dye contaminants into lower molecular weight organics and innocuous CO_2 and water. The results show that the commercial nano-TiO_2 is more efficient in decolourising the model dye and the textile effluents.

1. Introduction

In recent years, the photocatalytic decolourisation of dyes has attracted increasing attention due to prominent application prospects in the areas of treatment of pollutants in water [1,2,3]. Nanoparticles of titanium dioxide, nano-TiO_2 were evaluated for their ability to decompose organic contaminants [4].The nano-TiO_2 investigated includes a comparison of commercial, round Evonik P25® and particles synthesized via a hydrothermal method. Nanophotocatalysis using nanostructured semiconductors constitute one of the emerging technologies due to their high catalytic efficiency [5]. Photo catalysis has great potential for waste-water treatment. A universal range of semiconductors have been studied for their photocatalytic capacity and TiO_2 have been revealed to be the most active [6]. TiO_2 was selected as the photocatalyst for this work due to its high efficiency, low toxity and insolubility in water. Its operation at ambient temperature, low cost and it is ideal trace level pollutant removing ability [7] have also been considered. Because of its excellent photocatalytic activity, TiO_2 is an exclusive choice in many photo catalysis applications.
The photocatalytic decolourization of a dye is understood to take place according to the following mechanism in figure 1.

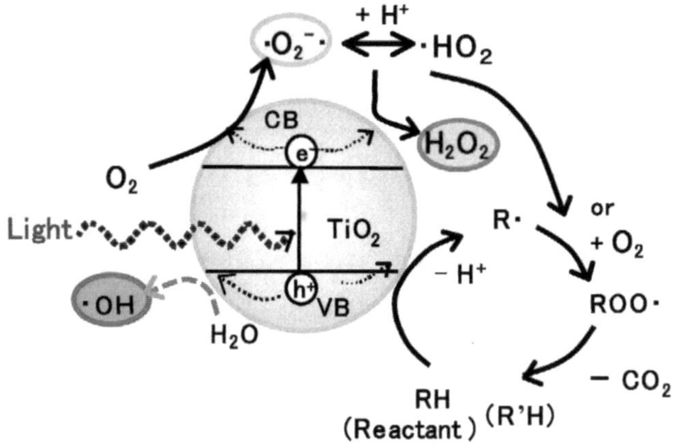

Figure 1. Showing photo catalysis mechanism [8].

As shown in figure 1, when the photocatalyst titanium dioxide (TiO_2) absorbs the ultraviolet (UV) radiation from daylight or other bright light source (fluorescent lamps), it produces pairs of electrons and holes. The electron of the valence band of the titanium dioxide becomes excited when illuminated by light The excess energy of this excited electron promoted the electron to the conduction band of titanium dioxide, therefore creating the negative-electron (e-) and the positive-hole (h+) pair. This stage is denoted as the semiconductor's 'photo-excitation' state [9].

The color transformation from dark-orange, (MO) to a transparent clear solution or a slightly colored solution is an indication of the decomposition of the organic dye contaminants into lower molecular weight organics and innocuous carbon dioxide and water. A number of methods were proposed to prepare nanostructure TiO_2 including microwave irradiation, chemical vapor deposition, spray pyrolysis, electrochemical deposition and sol–gel method. Photocatalytic oxidation process with metal oxides has a potential to break down the organic pollutants. Among the metal oxides, TiO_2 ranks among the most attractive and capable candidates for decolourisation of waste water from many industries [10-17] According to [18] recently nanostructured TiO_2-based photocatalytic process attracts due to its several advantages such as a high refraction index, biocompatibility, corrosion resistance, low cost, availability, non-toxic, ability to trace level pollutants and operation at ambient temperature [8]. Titanium dioxide (TiO_2) displays its exclusive photocatalytic activity at an excellent choice of a photo catalysis application [19]. Previously researchers have compared the photocatalytic activities of various TiO_2 powders using twelve types of commercial anatase TiO_2 and three types of rutile TiO_2, and concluded that the activity of the auto oxidation is much higher for anatase than that of rutile [20]. Despite the many known advantages of using TiO_2, it suffers from the shortcoming of having a large band gap (~3.2 eV) which restricts its use to the ultraviolet region. Suitable radiation wavelengths are not very abundant in the solar radiation spectrum that reaches the Earth, which limits the use of TiO_2 in solar energy utilization [21,22] observed a correlation between the photocatalytic activity in the UV photodecolourisation of phenol solutions and the particle size and that prismatic rutile particle with length around 5 nm and breadth of 60 to 100 nm showed the highest activity [22]. This is interesting as anatase is usually more reactive than rutile. The results were based on tests conducted on phenol solutions. None of these studies were scaled up to real industrial textile effluents, which will require ability to scale up the production of these novel shaped nanoparticles.

Your brightly colored clothes might have a dark story to tell. Hidden in the bright oranges (figure 2) of textile wastewater are some pretty harmful pollutants. Difficult to degrade, these compounds end up in waterways and may well outlast your outfit. Can we clean up and re-use water effluent from the textile industry [23].

Figure 2. Decolourisation of Methyl Orange at different photodecolourisation time intervals.

In this paper we attempt to tell a lighter side of the tale using special nano-based catalysts. Water polluted by textile dyes is a complex issue; impacting animal and plant life in waterways, and holds safety implications for human and agricultural purposes. For dry countries like South Africa under threat of climate change, these are important questions. The answer could lie in the use of light with a little help from nanosized titanium dioxide or rather, nano-TiO_2. Nano-TiO_2 acts as a photocatalyst with the ability to destroy organic contaminants from water[23]. Scientists know this process as photocatalytic treatment, harnessing the power of light. The photocatalytic properties of titanium dioxide have been known for decades and recently nanosized titanium dioxide has entered the market place. In this paper we are exploring these processes, and their application in treatment of dyes such as methyl orange etc.

The textile industry is one of the most chemically concentrated industries on earth and the top polluter of clean water (It takes about 2000 liters of water to produce enough fabric to cover one sofa). The clothing and textile industry is South Africa's sixth largest employer in the manufacturing sector and the 11th largest exporter of manufactured goods. The textile industry is not only amongst the largest industrial liquid waste generators, it is also chemically intensive. As a result, very large volumes of effluent containing a wide range of dyes, auxiliaries, salts, acids, alkalis and occasionally even heavy metals, are often generated [24]. The effluent from the dyeing is categorised by strong colour, high pH, high Biological Oxygen Demands (BOD), Total Suspended Solids (TSS), Chemical Oxygen Demands (COD), the presence of heavy metals like Cu, Hg, Cd, Zn, elevated temperatures and low or no biodegradability. Most dyes incorporated in the colour table are Azo dyes and available commercially, most of which are difficult to decolorize due to their complex aromatic molecular structure and synthetic origin [25].

The main aim of this paper is to compare the decolourisation of model dyes and textile effluent using both commercial and laboratory synthesized nano-TiO_2 under UV. In this study we investigated commercially available round shaped nano-TiO_2 and similar such particles synthesized via a hydrothermal method. The hydrothermal method of production produces rod and ribbon-shaped particles. The effect of these nanophotocatalysts was determined in laboratory-prepared solutions of model dye compounds. The sol-gel synthesis, although has high yield, suffers from a long ageing time.

2. Experimental details

2.1. Materials

Titanium dioxide (P25®, size 21 nm) as photocatalyst was supplied by Evonik. Methyl Orange (MO) as representative model dye compound was purchased from KIMIX, South Africa.

Synthesized nano-TiO_2 was synthesized by a sol gel technique using titanium isoproproxide as the titanium precursor, and 50 ml of HCI acid as the peptizing agent to control hydrolysis and favor rutile crystal structure growth. The pH of the dyes solution was adjusted with 1 M HCl. The titanium ion concentration was set at 0.5 M and the peptizing agent concentrations at 4 M and the synthesis was conducted over 24 hours. After the preparation of clear sol was aging by adding water drop-wise. Consequently, and after the aging, the complete solution including the gel is transferred to the hydrothermal reactor. It was hydrothermally treated at 220 °C for one hour.

2.2. Photo catalytic activity experiment

MO solution was prepared at 50 mg/l concentration amount of 0.04 g MO was added to 800 ml of deionized water in a one liter beaker separately. Then 2.4 grams of P25® TiO_2 was added and the solution was placed on a magnetic stirrer plate. The solution was sonicated for 30 minutes. Before starting running the experiment, a control sample of a 10 ml to aliquot was removed. A 9 W UV lamp was placed inside the solution to provide energy to the nano-TiO_2. The solution was irradiated for 30, 60, 120, 180, 300, and 420 minutes. After the solutions were exposed to UV light, the samples were centrifuged for 15 minutes in order to separate nano-TiO_2 from the rest of the solution before measuring the absorbance by a UV-Vis spectrophotometer.

2.3. Characterization

UV-Vis spectrophotometry (Perkin Elmer, Lambda 35) was used to measure the absorbance and % color changes of the dyes were calculated from the absorbance data. The structural properties of the prepared and commercial nano-TiO2 were analysed using X-Ray diffractometer. The particle shape and morphology were observed using High-Resolution Transmission Electron Microscopy (HRTEM). The surface area has been performed to indicate the sample surface area and porosity. Figure 3 below shows methyl orange calibration using UV-Vis spectrometer.

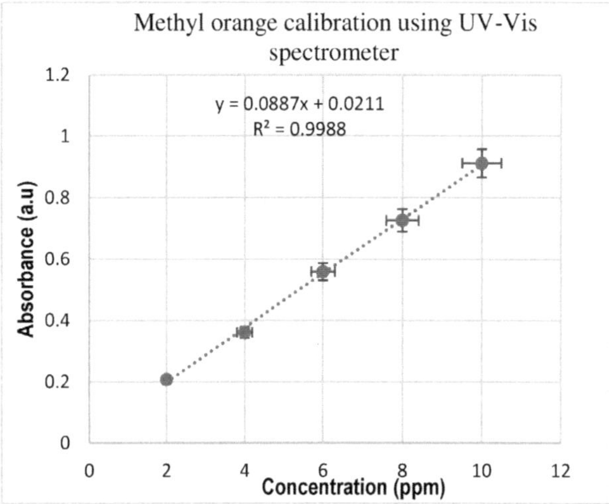

Figure 3. Methyl orange calibration plot.

3. Results and discussion

Titanium dioxide is considered for this treatment because of its availability, low cost and effectiveness as a catalyst, speeding up the removal of color from the textile wastewater. The proof is clear to see in images showing the decolourisation of methyl orange. UV irradiation of nano-TiO_2 excites electrons from a filled valence band to an empty conduction band, giving rise to electron-hole pairs. These generate free radicals which attack covalent bonds in the dyes. While nano-TiO_2 photo catalysis represents a striking method for breaking down these complex organic compounds to simpler compounds, scope remains for studies that can improve the efficiency of dye pollution. Measures such as these may yet lead to long-term solutions for treating textile industry wastewater.

3.1. UV-Vis analysis

Initially the control experiments were carried out under the following conditions: photolysis of dye solution with UV-light in the absence of TiO_2, dye solution with nano-TiO_2 in dark and the dye solution under irradiation of UV-light with TiO_2. After 30 min of magnetic stirring in the presence of photocatalyst without UV irradiation, more than 50%, 40% and 90% decrease in concentration of MO; the results of the measurements are shown in the Figure 4 - 5. This is due to the adsorption of dye molecule on the surface of nano-TiO_2. Figure 4 shows solutions of methyl orange (MO) after treatment with nano-TiO_2 under UV light at various time including: 0 min to 420 min, after centrifuging. The spectrum shows with a peak at approximately 462 nm, the decolourisation of MO at a different photodecolourisation times. As the color of the dyes becoming less concentrated the absorption peak intensity decreases with increasing decolourisation time which means that there are many molecules interacting with light. After exposing the MO solution for 420 min to UV irradiation, the reduction of

MO is up to 98% and the absorption peak is completely removed, the solutions has been decolorized. Figure 5, shows the decolourisation of MO suspension at a several decolourisation times. The color transformation from dark-orange to colorless MO demonstrates that the MO has been degraded by the TiO_2 nanoparticles under UV irradiation. Figure 6 shows MO treated by synthesized nano-TiO_2 is too longer to decolorize, only about 70% color removed after 420 min. The color transformation from dark-orange (MO) to a transparent clear solution or a slightly colored solution is an indication of the decomposition of the organic dye contaminants into lower molecular weight organics and innocuous CO_2 and water. Figure 7 shows the acual effluent from a textile company, the effluent took longer to decolorize because the high concentration dyestuffs preventing the transmission of UV light from the source through the solution. The effluent may contain other elements that could be interacting and or blocking light.

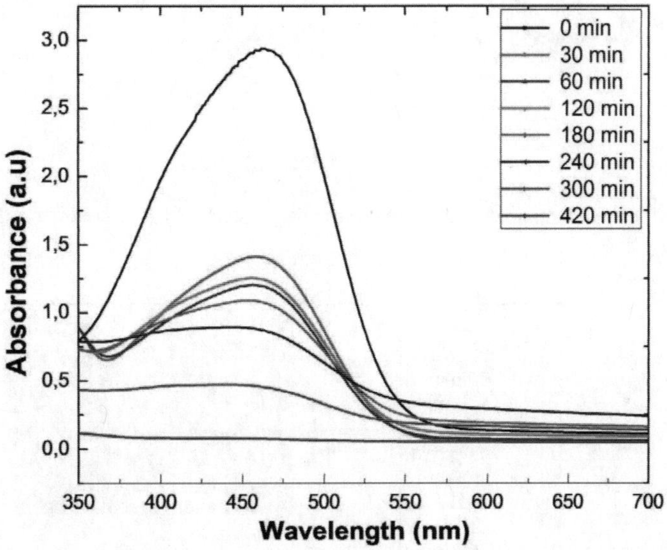

Figure 4. Showing absorption spectra of MO degraded by nano-TiO_2 under UV irradiation.

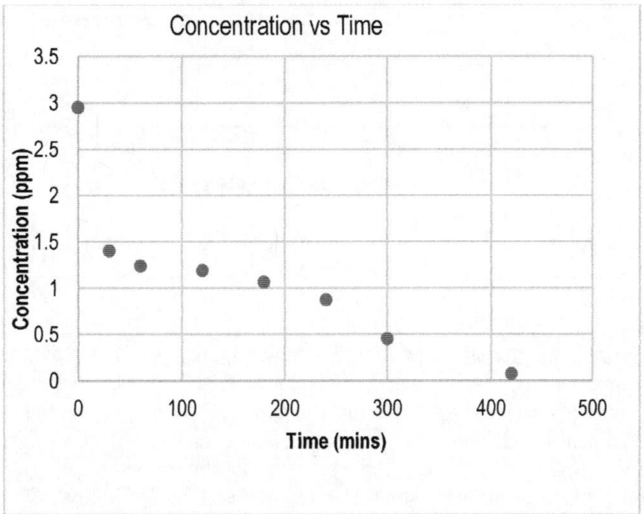

Figure 5. Showing plot of concentration versus decolourisation times.

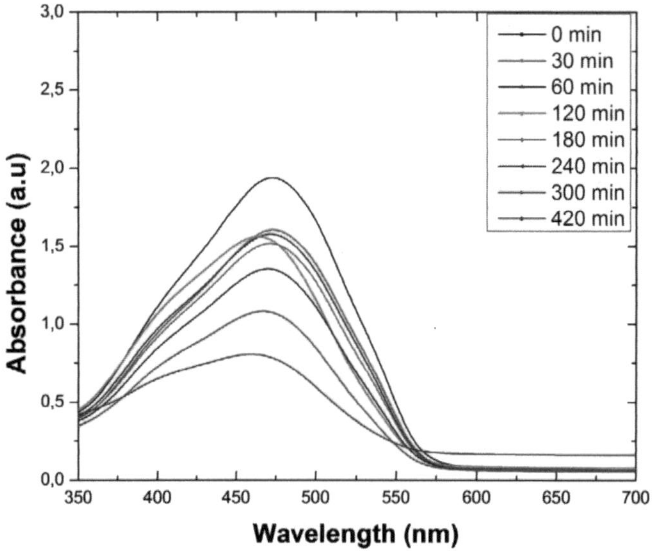

Figure 6. Showing absorption spectra of MO degraded by synthesized nano-TiO_2 under UV irradiation.

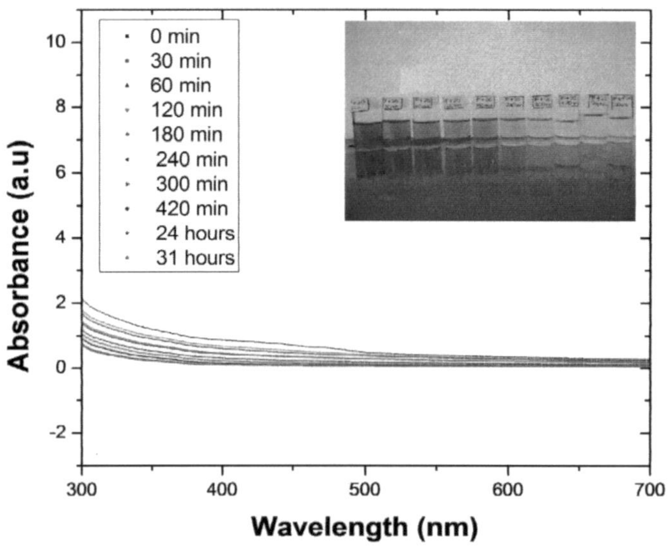

Figure 7. Showing absorption spectra of textile effluent degraded by commercial nano-TiO_2 under UV irradiation.

3.2. XRD analysis

Figure 8 below shows the characteristic peaks of the commercial nano-TiO_2 and synthesized nano-TiO_2. XRD spectrum of commercial nano-TiO_2 detected 10 characteristic peaks, since the commercial nano-TiO_2 is P25, it consists of anatase and rutile. So the 10 characteristic peaks detected correspond to both anatase and rutile. The peaks at 25.4° and 48.4° are the characteristic reflection for anatase and rutile [26,27]. The intensity of XRD peaks of the sample reflects that the formed nanoparticles are crystalline and broad diffraction peaks indicate very small size crystallite. Powder X-ray diffraction (XRD) was used for crystal phase identification and the estimation of the anatase–rutile ratio and the crystallite size of each phase present. In the case of synthesized nano-TiO_2 referred as NPs in the diffractogram below, six peaks were detected even though the peaks were not prominent like the ones for commercial nano-TiO_2 referred as power in the diffractgram. The synthesized nano-TiO_2 a prominent peak around 52. 4° which correspond to rutile.

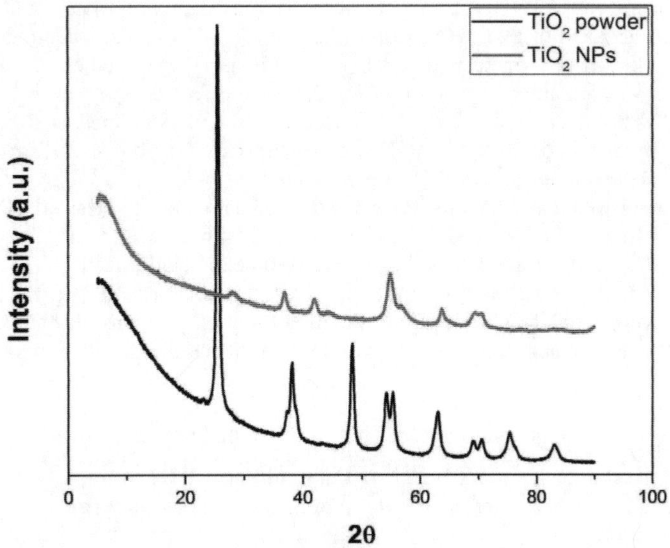

Figure 8. X-ray diffractograms of nano-TiO_2 powder in black line and synthesized nano-TiO_2 in red line.

3.3. HRTEM analysis

The morphology and particle size of commercial nano-TiO_2 and synthesized nano-TiO_2 were further investigated by HRTEM analysis.These images do not represent a novelty as TEM images of the popular P25 photocatalyst are frequently reported in the literature [28,29,30] It can be estimated that the particle size of synthesized nano-TiO_2 in figure 9 (a) and commercial nano-TiO_2 in figure 9 (b) nanoscale with the grain size from 20-30 nm range. From Figure 9 (a) and (b), HRTEM confirmed that both commercial nano-TiO_2 and synthesized nano-TiO_2 produced rod and ribbon-shaped particles.

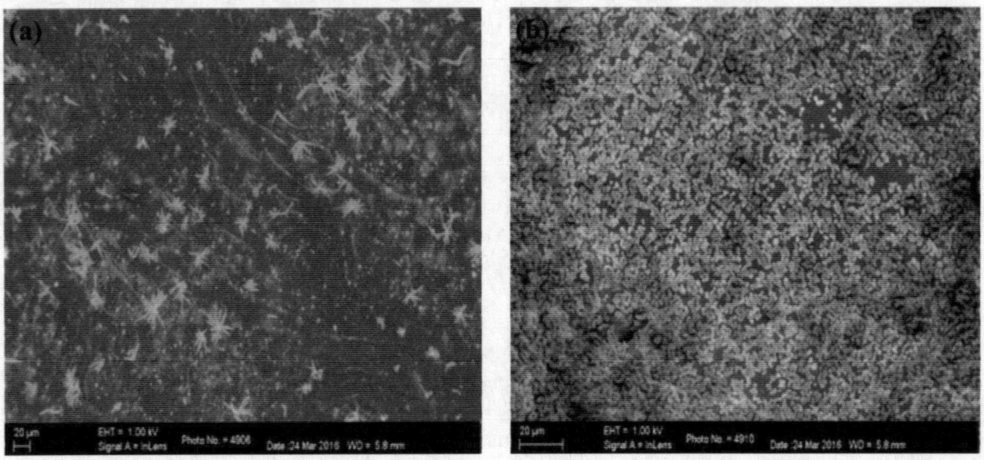

Figure 9. HRTEM photograph of the A (a) synthesized nano-TiO_2 (b) commercial nano-TiO_2.

3.4. BET Surface Areas and Porosity Measurements

The surface area has been performed to indicate the sample surface area and porosity of P25 TiO_2 and synthesized TiO_2. Surface area of the photocatalyst is a crucial factor in photocatalytic activity as it resolves the quantity of available catalytic sites. Specific surface area (SSA) measurements were carried out using BET analysis both P25 TiO_2 and synthesized TiO_2 as shown in Table 1 and Table 2.For the comparison, the surface area of P25 TiO_2 and synthesized TiO_2 was also acquired under the same experimental condition. It can be seen that the P25 TiO_2 has higher values of surface area than synthesized TiO_2. It was initially observed that P25 TiO_2 have a SSA of 85.90 m^2 g^{-1}, then the surface area increased to 102.26 m^2 g^{-1}. In the case of synthesized TiO_2, the SSA increased from 60.96 m^2 g^{-1} to 69.45 m^2 g-1.The corresponding Barrett–Joyner–Halenda (BJH) pore-size distribution and pore volume of P25 TiO_2 and synthesized TiO_2 are presented in table 1 and table 2 repectively. BJH adsorption pore size distribution with a mean values for P25 TiO_2 and synthesized TiO_2 are 23.64 nm and 31.17 nm. BET total surface area 375.42 m^2 g^{-1} and a total pore volume of 0.74 cm^3 g^{-1} for P25 TiO_2, the obtained BET total surface area and total pore volume for synthesized TiO_2 were 257.2 m^2 g^{-1} and 0.6 cm^3 g^{-1}. Due to the large density of crystalline TiO_2, this obtained BET surface area is very high for P25 TiO_2 and synthesized TiO_2 with high crystallinity [31-32]. As a result, the comparative high surface area and the assimilated porous structure will be valuable for the successive adsorption and photocatalytic reactions.

Single point surface area at P/Po = 0.196855792

Catalyst	BET SSA (m^2 g^{-1})	Pore volume (cm^3 g^{-1})	BJH average pore size (nm)
P25 TiO_2	85.90	0.22	9.91
	89.62	0.04	9.32
	97.64	0.24	8.39
	102.26	0.24	66.95

Table 1. Physical properties of the supported P25 TiO_2 photocatalysts.

Single point surface area at P/Po = 0.198604934

Catalyst	BET SSA (m^2 g^{-1})	Pore volume (cm^3 g^{-1})	BJH average pore size (nm)
Synthesized TiO_2	60.96	0.16	10.28
	63.12	0.06	10.30
	63.67	0.19	9.04
	69.45	0.19	95.06

Table 2. Physical properties of the supported synthesized TiO_2 photocatalysts.

4. Conclusion

From this paper the authors managed to decolorize MO and the actual textile effluent using nano-TiO_2 under UV irradiation. Although nano-TiO_2 photo catalysis is an attractive approach, there are some factors that restrain the performance of photo decolourisation such as catalyst activity -i.e. slow decolourisation rate of organics. In addition to a prompt removal of the colors, nanoparticles of titanium dioxide TiO_2 based photo catalysis were simultaneously able to fully oxidize the dyes, with a complete mineralization of carbon into CO_2. The resulting water is not only colorless but also detoxified when submitted to such a treatment. Photo catalysis appears as a valuable treatment for purification and reuse of colored aqueous effluents.

Acknowledgements

The author's would like to thank the Department of Science and Technology together with the National Research Foundation for funding this project, South Africa. We would like also to thank Mr Jules Lind and Prof Veruscha Fester from Cape Peninsula University of Technology, Chemical Engineering department, South Africa, for preparing TiO_2 nanoparticles and the CSIR for providing some of the resources for this project. Special thanks to Prof Janice Limson for valuable comments.

References

[1] Han F, Kambala V S R, Srinivasan M, Rajarathnam D and Naidu R 2009 *Appl. Catal.* A: *General* **359** 25
[2] Vinu R, Akki S U and Madras G 2010 *J. Hazard. Mater.* **176** 765
[3] Gaya U I and Abdullah A H 2008 *J. Photochem. Photobiol. C: Photochem. Reviews* **9** 1
[4] Nagaveni K, Sivalingam G, Hegde M.S and Madras G 2004 *Appl. Catal. B : Environ.* **48** 83
[5] Khataee A R, Vatanpour V and Ghadim A R A 2009 *J. Hazard. Mater.* **161** 1225
[6] Nickheslat A, Amin M M, Izanloo H, Fatehizadeh A and Mousavi S M 2013 *J. environ. public health* **2013** 9
[7] Joshi K M, Patil B N and Shrivastava V S 2011. *Arch. Appl. Sci. Res.* **3** 596
[8] Joshi K M and Shrivastava V S 2011 *Int. J. Environ. Sci.* **2** 8
[9] Bizani E, Fytianos K, Poulios I and Tsiridis V 2006 *J. Hazard. Mater.***136** 85
[10] Kawahara T, Miyazaki H, Karuppuchamy S, Matsui H, Ito M, Yoshihara M 2007 *Vacuum.* **81** 680
[11] Matsui H, Karuppuchamy S, Yamaguchi J and Yoshihara M 2007 *J. Photochem Photobiol A: Chem.* **189** 280
[12] Matsui H, Yamamoto S, Izawa Y, Karuppuchamy S and Yoshihara M 2007 *Mater. chem. phys.* **103** 127
[13] Matsui H, Kira K, Karuppuchamy S and Yoshihara M 2009 *Currt. Appl. Phys.* **9** 592
[14] Karuppuchamy S and Ito S 2008 *Vacuum* **82** 547
[15] Neppolian B, Choi HC, Sakthivel S, Arabindoo B and Murugesan V 2002 *Chemosphere* **46** 1173
[16] Sökmen M and Özkan A 2002 *J. Photochem. Photobiol. A: Chem.* **147** 77
[17] Muruganandham M and Swaminathan M 2004 *Sol. Energy Mater. Sol. Cells* **81** 439
[18] Santhi K, Manikandan P, Rani C and Karuppuchamy S 2015 *Appl Nanoscience* **5** 373
[19] Fujishima A, Hashimotoand K and Watanabe T 1999 Tokyo BKC 123
[20] Ray A K and Beenackers A A C M 1998 *Catal. today* **40** 73
[21] Fox M A and Dulay M T 1993 *Chemical reviews* **93** 341
[22] Testino A, Bellobono I R, Buscaglia V, Canevali C, D'Arienzo M, Polizzi S, Scotti R and Morazzoni F 2007 *J. Am. Chem. Soc.* **129** 3564
[23] Nyangiwe N 2014 **03,** npep.co.za.
[24] Barclay S and Buckley C 2000 **1**
[25] Özer A, Akkaya G and Turabik M 2006 *Dyes and pigments* **71** 83
[26] Ba-Abbad M M, Kadhum A A H, Mohamad A B, Takriff M S and Sopian K 2012 *Int. J. Electrochem. Sci.* **7** 4871
[27] Thamaphat K, Limsuwan P and Ngotawornchai B 2008 *J. Nat. Sci.* **42** 357
[28] Datye A K, Riegel G, Bolton J R, Huang M and Prairie M R 1995 *J. Solid State Chem.* **115** 236
[29] Martra G, 2000 *Appl. Cata. A: General* **200** 275
[30] Ohno T, Sarukawa K, Tokieda K and Matsumura M 2001 *J. Cata* **203** 82
[31] Zhang Y, Ying Zhang, Zhao Z, Chen J, Cheng L, Chang J, Sheng W, Hu C and Shunsheng C 2015 *Appl. Catal. B: Environ.* **165** 715
[32] Wang G, Zhu X and Yu J 2015 *J. Power Sources* **278** 344

Surface chemistry changes and microstructure evaluation of low density nanocluster polyethylene under natural weathering: A spectroscopic investigation

M Hamzah[1,2], M Khenfouch[3], A Rjeb[2], S Sayouri[2], D S Houssaini[2,4], M Darhouri[2] and VV Srinivasu[1]

[1] Chair for Superconductivity Technology, Department of Physics, College of Science, Engineering and Technology, Florida campus, University of South Africa, Cnr Christiaan de Wet & Pioneer Avenue Florida 1709, Johannesburg, South Africa
[2] LPTA, Faculty of Sciences Dhar el Mahraz, University Sidi Mohammed Ben Abdellah, Laboratory of Theorical and applied Physics LPTA, PO Box 1796 Atlas Fez 30 000, Morocco
[3] Africa Graphene Center, Physics department, Eureka building, College of Science, Engineering and Technology, Florida campus, University of South Africa, Cnr Christiaan de Wet & Pioneer Avenue Florida 1709, Johannesburg, South Africa
[4] Department of Industrial Engineering, Faculty of Sciences and Technicques, University Sidi Mohammed Ben Abdellah, BP 2202 Fez, Morocco
Corresponding author: hamzah.13.mohammed@gmail.com

Abstract. Polyethylene is the most commonly used plastic in daily life, covering wide areas of application e.g. this polymer is used as a greenhouses covering material. This article investigates the effect of photo-oxidation on commercial unstabilised Low Density Polyethylene (uLDPE), as result of outdoor weathering factors. In this study, the samples were exposed for four months to the natural weather. The physico-chemical effects of natural ageing were studied by attenuated total reflection Fourier transform infrared (ATR-FTIR) and X-ray photoelectron (XPS) spectroscopy to elucidate the chemical composition, the nature of chemical bonds established and further to interrogate the changes that occur on the surface of the uLDPE samples. The main chemical change of uLDPE results in the formation of different kinds of carbonyl and vinyl groups identifiable in the ATR-FTIR and XPS spectra. The degree of crystallinity for these samples was calculated in terms of time exposure. An increase in the degree of crystallinity due to chemicrystallization was observed, which we indicative of the occurrences of chain scission. During outdoor exposure it was found that the photo-oxidation results in the formation of chain scission occurrences via Norrish type II reactions.

1. Introduction

For many decades, polymers have been increasingly used across various fields in divers sectors.Today, the areas of application are still in continuous expansion due to the wide range of uses that polymers serve in commercial; and industrial markets. The choice of using a polymer instead is often motivated by cost reduction in conjunction with a reduction in weight, without sacrificing performance. Amongst synthetic polymers, polyethylene and polypropylene are extensively used in the polymer industry and alone represent more than 50% of the total volume of the thermoplastics manufactured in the world. In view of the importance of polyethylene, the study of its performance characteristics have been carried out for many years.

Polyethylene is the polymer with the largest production and commercial uptake within the sector. Low density polyethylene (LDPE) is particularly well known for its use in many outdoor applications, especially for greenhouses, as covering material. Unfortunately, the polymers (polyolefins) are subject

to natural (outdoor) weathering, which lead to material degradation. Weathering is particularly severe for organic materials, because it combines the photophysical and photochemical effects of ultraviolet (UV) radiation, with oxidative effects of atmospheric oxygen and the hydrolytic effects of water [1]. Weathering is commonly defined as the undesirable change produced by outdoor exposure [1]. The study of the aging of polyethylene through weathering is of great importance for the development of more resistant materials, in comparison to what was previously available. For this reason, many studies have been conducted over the years, for instance, to investigate the natural aging of polyethylene [2-11], and to study the artificial aging [12-16].
The photodegradation of polymers is mainly caused by the introduction of chromophores, such as catalyst residues, hydroperoxide groups, carbonyl groups, and double bonds, during polymer manufacture [17]. Studies on the luminescence polymers have expanded the field of knowledge about the absorption of molecules and, therefore, on photooxidation mechanisms. Charles by et al. [18] were the first to report the phosphorescent emission of polymers containing carbonyl impurities. Allen et al. [19] were interested by the phosphorescence of species formed during the thermal oxidation and photo-oxidation process for commercial polyolefins. They observed the presence of species having an excitation wavelength in UV-B range. These authors attribute this phosphorescence to ketones and aldehydes which would, according to them, the main initiators of photo-oxidation [19]. Photochemical initiation (photolysis ketones) is responsible for the formation of various oxidation products, the photolysis of ketones proceed according Norrish I and Norrish II reactions.
Polyethylene is a semi-crystalline polymer, the most stable crystal structure of polyethylene and most commonly encountered corresponds to an orthorhombic crystal lattice [20]. On a morphological point of view, polyethylene can be considered as alternating crystalline and amorphous layers (two-phase system). Oxidation damage confined to the amorphous phase causes the break of the tie molecules that link the various crystalline lamellae, and lead to a sudden loss of microstructural properties at break [21]. The weathering of polyethylene affects the crystallinity. The shorter chains produced during chain scission are more mobile and are able to crystallize more readily, which increases crystallization and associated embrittlement [17, 22].
The current work presents the analysis of natural ageing behavior of uLDPE samples. Aging characterization was performed by attenuated total reflection Fourier transform infrared (ATR-FTIR) and X-ray photoelectron (XPS) spectroscopy. In this study we discussed the contribution of photolysis ketones reactions according Norrish I and II, we used ATR-FTIR spectroscopy to monitor changes in crystallinity during natural weathering. changes Complementarily to infrared spectroscopy, the aging behavior was investigated by using XPS spectroscopy, which is useful to study the determination of atomic composition of the surface of aged unstabilised polyethylene, therefore the level of oxidation of the surface may be assessed. Furthermore, and by means of the deconvolution of the carbon and oxygen, the XPS spectra indicate the variety of the bonds.

2. Experimental

2.1. Materials
The commercial uLDPE agro films was kindly supplied by the Algerian company ENIP (Entreprise Nationale des Industries Petrochimiques) without additives, having as trademark: PEBD B21 ENIP. The density and the melt flow index given by the supplier are respectively d=0,923 g/cm^3 and 0,019 g/min. The samples were about 20x20 mm. The unaged sample was analyzed by ATR-FTIR and XPS spectroscopy and served as reference.

2.2. Natural (outdoor) weathering
The samples uLDPE have been weathered under natural sunlight (days and nights) for four months during the period from July to October. Natural aging of these samples was carried out at the roof of a building in Fes city, Morocco. The average temperature varied between 18 and 34°C and the daily average was 26°C, the relative humidity average was 41%. This period was chosen because the climate is hot and dry. The samples were then placed on wooden board (100x100 cm), the board was then placed on roof top facing southwards at 45 angle with the roof or plane of the earth in accordance with NF T-51165. The samples directly exposed to the weather were taken out monthly and characterized regularly by ATR-FTIR and XPS spectroscopy.

2.3. Material characterization

2.3.1. Attenuated total reflection Fourier transform infrared spectroscopy
ATR-FTIR spectra were recorded on a Magna-IR 560 using ZnSe crystal. The spectra were recorded in the 400-4000 cm^{-1} range, with a resolution of 4 cm^{-1} and scan number of 400. All spectra were taken in logarithmic absorbance mode. Experiments are performed applying the ATR equipment with the FTIR spectrometer. The sample surface exposed to photo-oxidative conditions has been in contact with ZnSe crystal.
The spectral bands of the investigated functional groups of the polymer were normalized with respect to a reference band and internal standard. Obviously, the reference band was present in unstabilised polyethylene samples and its intensity and wavelength did not change upon the aging time. The band characteristic of the wagging deformation of methylene groups (1368 cm^{-1}) was accepted as the reference band [13,23,24]. The relative intensity of a specific absorption band was calculated as the ratio of the surface area of this band to that of the reference band.
Infrared ATR spectroscopy provides a qualitative and quantitative analysis of the chemical functions present in the surface. The qualitative analysis was based on separation of the peaks by means of a Lorentzian fitting of the infrared spectra in the region 1600-1800cm^{-1} and 1000-1250 cm^{-1} using OriginPro8 software. The quantitative analysis was performed using the peak surface areas. The quantification of the different photo-oxidation products was carried out by means of calculation of carbonyl and vinyl index according to the following equations:

$$I_{carbonyl} = \frac{A_{C=O}}{A_{1368}} \quad (1)$$

$$I_{vinyl} = \frac{A_{C=C}}{A_{1368}} \quad (2)$$

Where $A_{C=O}$, A_{909} and A_{1368} are respectively the areas peaks of carbonyl group (1670-1800 cm^{-1}), the area peaks of vinyl group 909 cm^{-1}. These indexes were normalized using the peak at 1368 cm^{-1}.
The study of the crystallinity changes for orthorhombic uLDPE after weathering has been determined by the infrared spectroscopy. The rate of crystallinie content X_{cr} was calculated via the empirical equation of Zerbi et al. [25]:

$$X_{cr} = 100 - \frac{((1 - I_a/I_b)/1.2333)}{(1 + I_a/I_b)} 100 \quad (3)$$

Where X_{cr} percentage of crystalline content, I_a represents the area peaks corresponding to polyethylene crystalline content (729 cm^{-1} or 1472 cm^{-1}), and I_b represents the area peaks corresponding to the amorphous content (719cm^{-1} or 1462 cm^{-1}). The constant 1.2333 corresponds to the relation of the intensities of these bands for a completely crystalline polyethylene.

2.3.2. X-ray photoelectron spectroscopy(XPS)
The XPS set-up used was a Kratos Axis Ultra system with an electrostatic-magnetic hybrid lens. The sample is fixed on a stainless steel sample holder, using stainless steel screws also. A surface of about 400x800 μm was analyzed by a monochromatic Al Kα source at incidence angle of 30°. The base pressure analysis chamber is $8x10^{-10}$Torr. When recording data, the pressure rises to a value of $5x10^{-9}$ Torr. During the analysis of peaks, we used Shirley background, and for all spectra, the Gaussian/Lorentzian ratio was 40. Calibration of XPS survey spectra were based on the C1s peak, while the calibration of C1s spectra was based on the peak Csp3 and calibration of O1s spectra of the peak C-OH.
For the unaged sample, the XPS analysis was performed without cleaning. For the aged samples, there has been cleaning with distilled water and drying with compressed air before the measurements.

3. Results and discussion

3.1. Attenuated total reflection Fourier transform infrared spectroscopy

3.1.1. Structural modifications and chemical changes
The usual bands of pristine uLDPE are located in the regions: 3000-2800, 1550-1400 and 750-650 cm^{-1} [26], and are illustrated in figure 1, we are in the presence of three strong absorption bands.

1) The two peaks of the doublet 2846 and 2914 cm^{-1}, with a very strong intensity, correspond to the vibration mode of the chain CH_2 groups [7, 14], respectively assigned to the methylene symmetric and asymmetric C-H stretching vibration [26-28].
2) The two peaks of the doublet 1462 and 1472 cm^{-1}, with a strong intensity, are ascribed respectively to the bending deformation of methylene CH_2 groups in amorphous and crystalline domains respectively [29,30].

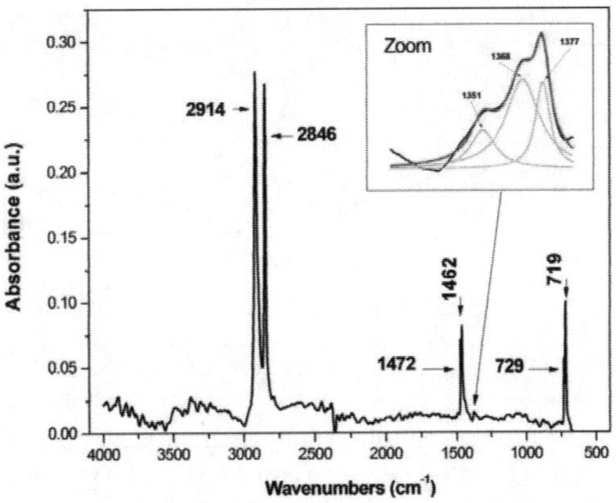

Figure 1. ATR-FTIR spectrum of the non-aged uLDPE.

3) The two peaks of the doublet 719 and 729 cm^{-1}, with a strong intensity, are assigned to CH_2 rocking deformation [26, 30-32] in amorphous and crystalline domains respectively [27, 30].
In addition to these three usual bands above, there are also three peaks that have low intensity in the region 1390-1320 cm^{-1} (figure 1). The peak in 1377 cm^{-1} ascribed to CH_3 symmetric deformation [26, 31], the peak 1368 cm^{-1} and 1351 cm^{-1} inset are assigned both to the wagging deformation CH_2 group [13, 23,24,26].
After defining characteristic bands of the virgin uLDPE sample that have not undergone the natural aging (figure 1), we have followed the progress of degradation, with the exposing time, of uLDPE samples under the influence of weather (natural climatic conditions). Due to aging, new peaks appear alongside usual bands characterizing the C-H bond,-CH_2-, and CH_3 groups (figure 2); we define three regions containing different bands that varied clearly upon the natural aging (figure 3A, 3B). We will discuss simultaneously the peaks that reveal these regions.

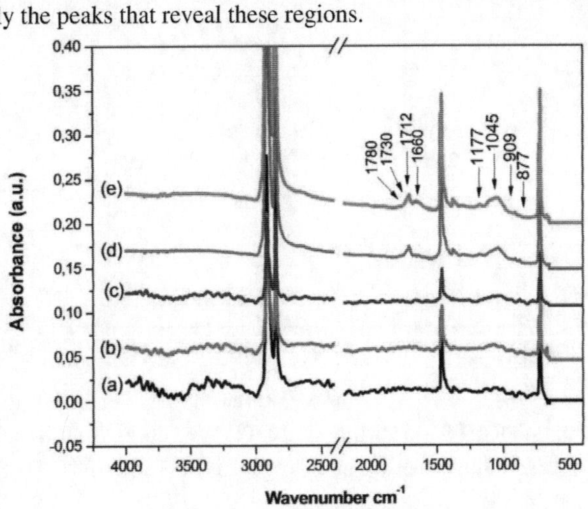

Figure 2. ATR-FTIR spectra of uLDPE for different times of exposure to natural aging.
(a) non-aged, (b) aged one month, (c) aged two months, (d) aged three months, (e) aged four months.

In the region 850-1000 cm^{-1} (figure 3A), we observe after two months of exposure, the emergence of a small peak of low intensity to 877 cm^{-1} and one month later (sample d), next to the latter peak appear a second peak at 909 cm^{-1} with similar intensity. In the final stage of aging (sample e), we find that the small peak at 909 cm^{-1} persists and the other around 877 cm^{-1} disappears (figure 3A). The peaks at 877 and 909 cm^{-1} correspond to vinylidene (-CH=CR_1R_2) and vinyl (-CH=CH_2) respectively [3, 23,33]. The consumption of the vinylidene group is known to be due to their reaction with alkyl radicals (RH·) which leads to the formation of crosslinks [21]. These results are in perfect agreement with those established by Chabira et al. [21].

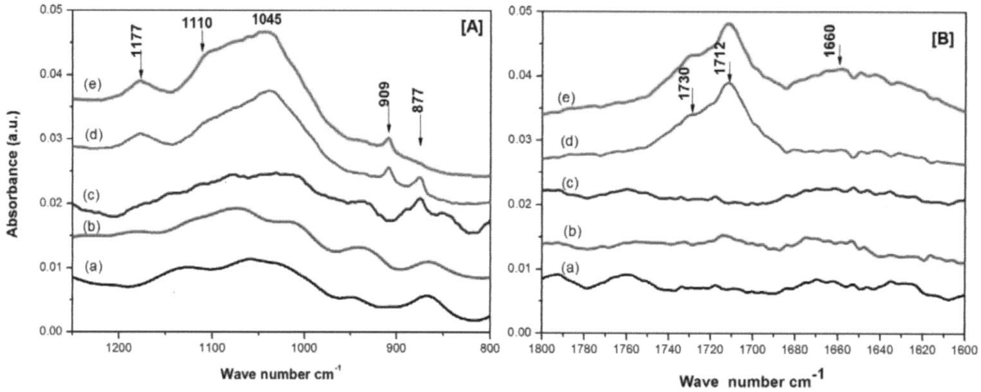

Figure 3. ATR-FTIR spectra of uLDPE samples for different times of exposure to natural aging. **(A)** In the region 850-1250- cm^{-1}, and **(B)** In the region 1600-1800 cm^{-1}, (a) non-aged, (b) aged one month, (c) aged two months, (d) aged three months, (e) aged four months.

In the region 1000-1250 cm^{-1} (figure 3A), the first signs of the aging effect appear in this region, which is known as an amorphous band [5]. Amorphous regions are prone to oxidative chain scission [35]. After two months of exposure appears a peak around 1031 cm^{-1} related to the formation of the C-O bond due to oxidation processes. This latter peak after three and four months shift to 1040 cm^{-1} and about 1045 cm^{-1} respectively (figure 3A).

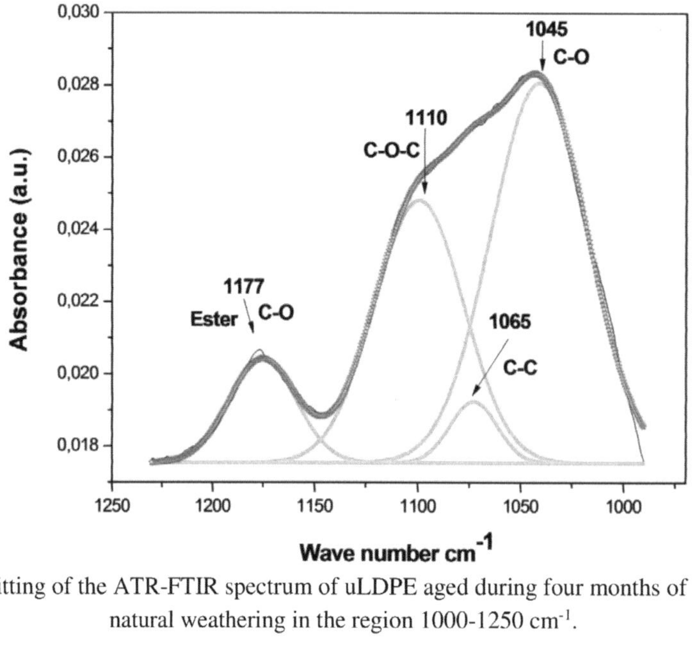

Figure 4. Fitting of the ATR-FTIR spectrum of uLDPE aged during four months of exposure to natural weathering in the region 1000-1250 cm^{-1}.

The diffusion of oxygen through the amorphous regions is much easier than in the crystalline regions [37]. Oxidation damage confined to the amorphous phase causes the break of the tie molecules that link the various crystalline lamellae [21]. In the third month, near the peak at 1040 cm^{-1} rises two peaks 1110 and 1177 cm^{-1} with small intensity (figure 3A). In the last stage of aging, we have performed a Lorentzian fitting to separate and identify all the peaks (figure 4). We have identified four peaks 1045, 1065, 1110 and 1177 cm^{-1} are ascribed respectively to the stretching vibration of CH_2-O group [7,14], the stretching vibration of the C-C bond in the amorphous phase of the polyethylene [36], the stretching vibration of the C-O bond due to the ether group C-O-C [27], and the latter peak is assigned to the stretching vibration of the C-O bond characteristic of the ester group [27], which is also defined by the C=O bond that we will see in what follows.
In the region 1600-1800 cm^{-1} (figure 3B), until the second month we have a complete absence of peaks. In the third month of aging (samples d), the 1712 and 1730 cm^{-1} are correlated with the presence of carbonyl absorption band [37]. One month after (samples e) the vinyl group at 1660 cm^{-1} appears in the vicinity of the carbonyl groups and is assigned to C=C stretching vibration [27]. The formation of carbonyl groups and vinyl groups can be interpreted as a direct indication of main chain scission [37,38].

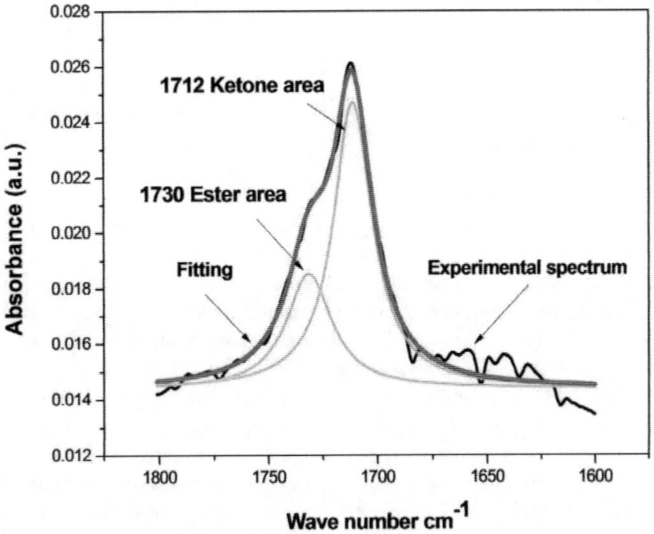

Figure 5. Fitting of the ATR-FTIR spectrum of uLDPE aged during three months of exposure to natural weathering in the region 1600-1800 cm^{-1}

In order to highlight the nature of chemical bonds between atoms (functional group), and also to identify and monitor the carbonyl band which is an entanglement of compounds, a Lorentzian fitting was performed for the last two months of natural aging. The figure 5 shows that after three months of exposure there is no trace of vinyl group, no peaks in the 1600-1670 cm^{-1}. In the carbonyl region 1670-1800 cm^{-1} we distinguish two carbonyl compounds, 1712 cm^{-1} is assigned to ketone group, and 1730 cm^{-1} is belonged to aliphatic ester group [27,34,39] both carbonyl groups are ascribed to the stretching vibration of C=O bond. According to Barbara Stuart [27], the aliphatic ester group has two polar bonds C = O and C-O respectively in the bands 1750-1730 and 1300-1100 cm^{-1}. Indeed, we have defined previously the stretching vibration of the C-O bond at 1177 cm^{-1}characteristics of the aliphatic ester (figure 3A and 4). This result is more or less similar to that evoked by Yang et al. [40] who had attributed the ester group to the peaks 1741 and 1178 cm^{-1}. Furthermore, Tijani [37], in his study of linear low density polyethylene under different natural and accelerated weathering conditions, assigned the peak at 1735 cm^{-1} to esters, and noted that the only apparent difference observed compared to spectra obtained in accelerated aging, is related to the esters groups: the shoulder at 1735 cm^{-1} have a lower intensity in natural aging. In the last stage of natural aging, by means of fitting (figure 6) in addition to ketone and ester groups, a small peaks of γ-lactone (1780 cm^{-1}) and vinyl groups (1660 cm^{-1}) appears. The fitting also exhibit that after four months of exposure, the area of ketone group is smaller than that of ester group (figure 6). This result can be due to the formation of the vinyl group which derived from the photolysis of the ketone according to Norrish II reaction.

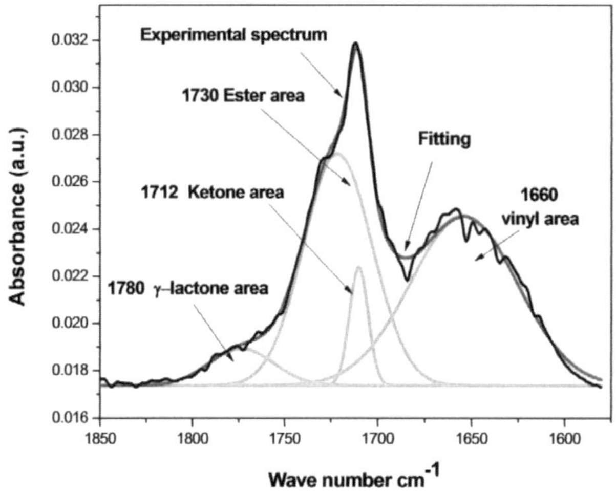

Figure 6. Fitting of the ATR-FTIR spectrum of uLDPE aged during four months of exposure to natural weathering in the region 1600-1800 cm^{-1}

3.1.2. Quantification of the photo-oxidative products

The study of the evolution of vinyl and carbonyl degradation products gives a good indication of the progress of the polymers aging. In particular, some studies have shown that the amount of carbonyl can be directly correlated with the degradation of the polyethylene [41-43]. After having identified the carbonyl groups via fitting, we focus on the quantification of carbonyl and vinyl degradation products. The measurement of vinyl and carbonyl groups formation involves the calculation of the indexes previously defined I_{Vinyl} and $I_{carbonyl}$. Table 1 gives the evolution of these components, includes three carbonyl species and shows their evolution over time. We find that there is nothing to report for the first and second months of exposure; the absence of carbonyl massif is also clearly seen in the figure 3B. In the third month the amount of ketone is quite important, the ester group is also present, but with a smaller amount compare to this of ketone. At this stage the ketone index is almost double that of the ester. After three months of exposure γ-lactone appears and its index was calculated to be 0.62, and the vinyl index increases. Moreover, we note that the index for the ester increases linearly until a maximum value in the fourth month, however, the index of ketone decreases.

Table 1. Carbonyl and vinyl index depending on the exposure time of uLDPE to natural aging.

Natural aging time	**carbonyl Index**			**vinyl Index**
	Ketone Index	Ester Index	γ-lactone Index	Vinyl index
One month	-	-	-	-
Two months	-	-	-	-
Three months	2,54	1,36	-	0,12
Four months	0,57	4,14	0,62	4,52

3.1.3. Microstructural evaluation: crystallinity changes

The increase in carbonyl and vinyl groups formation in polyethylene after weathering is known to be proportional to the number of chain scissions that occur in the polyethylene [17,22]. Effect of chain scissions induced crystallinity modification. Therefore, another way for characterizing degradation and

following the microstructural changes after weathering is to monitor the crystallinity variation of polyethylene samples, which is a good indication of the chain scissions mechanism. Moreover, the mechanical behavior of semi-crystalline polymers, such as low density polyethylene, is directly dependent on the microstructure. Different parameters can influence the mechanical behavior of these materials such as the rate of crystallinity or the size and distribution of crystallites [44].
Table 2 illustrates the variation of the crystalline content of LDPE samples versus the natural aging time via the empirical equation of Zerbi et al. [25] defined before on equation (3). We can notice a considerable difference in terms of selected bands (table 2). The intensity ratio 729 cm^{-1}/719 cm^{-1} leads to a crystalline content ranging from 59.60% to 66.81%, and the bands 1472 cm^{-1}/1462 cm^{-1} leads to crystallinity values between 53.31% and 55.08%. We find that the values of crystallinity, for the band 1472 and 1462 cm^{-1}, are versatile. Kaci et al. [45] have studied the crystallinity changes of low density polyethylene films exposed to natural weathering using FTIR and DSC and WAXS analyses. They concluded that the measurement of crystallinity through the bands 1474 and 1464 cm^{-1} attributed to scissoring modes leads to unreliable results. Thus, we will adopt the crystallinity values calculated from the methylene rocking bands at 729 cm^{-1} and 719 cm^{-1}.

Table 2. Crystalline and amorphous contents evaluated by means of Zerbi et al. Equation using ATR-FTIR spectroscopy.

Exposure Time (months)	Amorphous content (%)	Crystalline content (%)	I_a Value	I_b Value
		Bande spectrale I_a = 729 cm^{-1} et I_b=719cm^{-1}		
0	40.40	59.60	1.0315	0.3454
3	35.64	64.36	1.323	0.515
4	33.19	66.81	1.4827	0.6212
		Bande spectrale I_a = 1472 cm^{-1} et I_b=1462 cm^{-1}		
0	46.69	53.31	0.9793	0.2635
3	46.63	53.37	1.4953	0.4033
4	44.92	44.08	1.7439	0.5

The behavior of crystallinity shown in table 2 indicates ascension in the value of the crystalline content from 59.60% for pristine sample, to 64.36%, after three months of exposure. The crystalline content continues to increase to reach 66.81% after four months of exposure, which means an increase of 7.21% due to natural aging of uLDPE. Mendes et al. [46] reported by using thermal analysis DSC, in the case of non-stabilised hight density polyethylene HDPE submitted to a natural aging in Rio de Janeiro, that after 74 days of exposure, they marked 6% increase in crystallinity. Kaci et al. [45] studied the crystallinity of unstabilised low-density polyethylene (LDPE) films under natural weathering conditions in Bejaia Mediterranean town of Algeria, via FTIR spectroscopy, the value of the crystallinity content have increased by 7% after 170 days of exposure. The crystallinity measurements depend on the geographical site which they are submitted, and we can say that for some polyolefins, crystallinity increases with natural weathering exposure.
An increase in crystallinity can also be used as an index of the degradation of polyethylene via chain scissions occurring mainly in the amorphous phase. These latter reduce the density of entanglements in the amorphous phase, which allows shorter molecules to crystallize because of their higher mobility [17,33]. Therefore, the secondary crystallization, takes place in the amorphous phase of the polyethylene. The amorphous phase of the material is characterized by a free portion containing entanglements and a part linked to crystalline lamellae.

3.1.4. Mechanisms of photo-oxidation reactions

Photo-oxidation aging develops under the joint action of ultraviolet light and oxygen, making it a leading cause of the polymers aging. It belongs to the class of chemical aging associated with oxidation phenomena; the action of light produces radicals which subsequently react with oxygen. The absorption of UV solar radiation and the ability of oxygen to diffuse into the material have a crucial role.

The solar radiations reaching at the surface of the earth called UV-B (280-315 nm) are responsible for the majority of polymer degradation [47]. Thereby, Allen et al. [48] suggest that the polyolefins should not be affected by solar radiation. Likewise Gijsman et al. [49] confirm this, and claim that the polyolefins which belongs polyethylene only absorb in the deep UV ($\lambda<200$ nm). On the other hand, relying on the absorption of thin films of polymers (0.05 mm thickness) in the UV (220-390 nm), Searle [50] specifies that the aliphatic polymers such as polyethylene and polyvinyl chloride are not able to absorb UV solar radiation reaching the earth, so saturated in the area 280-400 nm. The main absorption band of these polymers would be located, according to the authors, below 220 nm. Guadagno et al. [41] concede the photochemical aging to the absorption of solar radiation by the impurities present in the polymer.

Studies conducted on the polyolefins by luminescence have expanded the field of knowledge about the absorption of molecules and, therefore, on photo-oxidation mechanisms. Charles by et al. [51] were the first to report the phosphorescent emission of polyolefins containing carbonyl impurities. Allen et al. [19] have studied the phosphorescence of species formed during the thermal oxidation and photo-oxidation process for commercial polyolefins. They observed the presence of species having an excitation wavelength higher than 290 nm. These authors attribute this phosphorescence to ketones and aldehydes which would, according them, the main initiators of photo-oxidation [19]. Therefore, according to these authors, and to our results, we can conclude that the absorption of sunlight, for our samples, may be derived from chromophore groups such as the catalytic residues and carbonyl impurities (ketones) formed during the polymerization phase. Ketones absorption bands would be between 330-220 nm and below 340 nm, and then contribute to the initiation of photo-oxidation. Thus, photolysis of ketones seems to be explained preferentially by Norrish I and II reactions.

We discuss, in what follows, the contribution of ketones photolysis reactions according Norrish I and II. The Mechanism of the photolysis of ketone according Norrish I is a free radical process, leading both to chain scissions and crosslinking, while the photolysis by Norrish II is not a radical process and leads only to chain scissions, accompanied by the production of vinyl groups. The formation of unsaturated vinyl at 909 and 1660 cm^{-1} (figure 3 and 6) indicates that we are in the presence of Norrish II reaction. In other words, stopping the formation of vinyl attest another way by photolysis of ketones according Norrish I. Thereby, the transition from Norrish II to Norrish I can not occur, since we notice that after three months of exposure, vinyl concentrations always continue to evolve (figure 7 and table 1). On the other hand, the passage of the photolysis of ketones from Norrish II to Norrish I first requires a deceleration of reaction rate of vinyl formation followed by a cessation of the formation of unsaturated vinyl; condition not satisfied in our spectra.

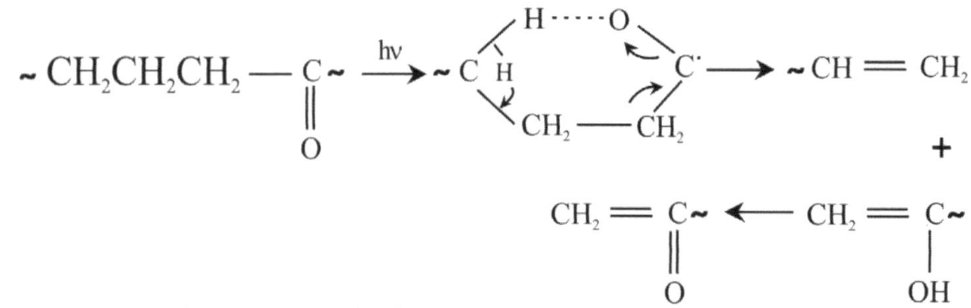

Figure 7. Photolysis of ketones according Norrish II mechanism.

In order to rule on the predominance between Norrish I and II, Carlsson and Wiles [52] studied the photolysis of models ketones: ketone A in mid-chains and ketone Bat the end of chains. They concluded that ketone A is photolyzed primarily by a Norrish type I scission to give carbon monoxide and two macroradicals. Whereas ketone B is photolyzed by a Norrish type II scission to give Ketone

and an unsaturated polymer chain end. This proves again that the mechanism of photo-oxidation of our samples is made according to Norrish II process, and that it is ketones at the end of chains.
After discussing the predominance of reactions Norrish I and II one over the other, we are looking now to focus our study on the consequences of these reactions. The effects of photo-oxidation on macromolecular scale could be summed up in the competition of two phenomena: the formation of chain scission and crosslinking. The kinetic of these two phenomena is governed by Norrish I and II mechanisms. The crosslinking reactions derived from Norrish I process are in reality bridges of dioxygen formed by reaction between peroxyl radical and vinyl. Thus these possible reactions also could explain the decrease in vinyl concentration. Indeed, Pagès et al. [53] observed by IR analysis, after 60 days of HDPE exposure to natural aging, a decrease of absorbance vinyl groups they attribute to crosslinking. This case is not convenient to our results (table 1), the vinyl index increase with exposure time, and consequently the mechanism of photo-oxidation follows Norrish II process. Chain scissions reactions could arise from β scission of alkoxyl radical [54-56], and from photolysis of ketones according Norrish type I and II. A production of chain scission through β scission of alkoxyl radical is excluded, since we have no traces of aldehydes and acids in our spectra (figure 6). In the second case, the formation of chains scission according Norrish I is marginalized, as long as vinyl groups are still present in the final stages of aging. So we are left with the possibility of the origin of the chain scission mainly occurs through photolysis of ketones according Norrish II mechanism.

3.2. X-ray photoelectron spectroscopy

The XPS surveys of uLDPE samples in figure 8 are used to identify the kinds of atoms present in the surface. Figure 8 shows the presence of an intense peak with the maximum at 285 eV assigned to the carbon atoms C1s, that is the main constituent on the surface of the unaged samples and for the others polyethylene samples aged under natural weathering during four months. Oxygen (O1s at 532 eV) is the major contaminant compared to other surface elements, and the intensity of the corresponding peak increases during the aging. Minor contaminants are also found, mainly small peaks of silicon (Si2p at 102 eV and Si2s at 153 eV), and aluminum (Al2p at 75 eV), we can attributed these contaminants to the clays which can be carried by the wind and settle on the surface of our samples during natural aging.

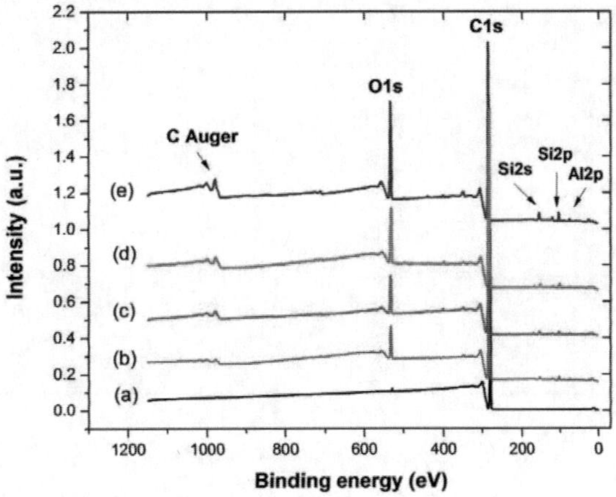

Figure 8. XPS survey spectra of uLDPE samples. (a) non-aged; (b) aged during one month; (c) two months; (d) three months and (e) four months.

Especially, illite is dominant in Fes (northern Morocco) clays, but is associated with kaolinite [57]. A peak of carbon Auger is also present at 986 eV, obviously, hydrogen was also present in the samples but it is undetectable with the XPS technique. The characteristic feature of the recorded survey scans is that the intensity of the O1s peak increases with exposure time and that of carbon decreases (figure 8 and figure 9).
Table 3 gives the percentages of atomic concentrations of the chemical species present on polymer surfaces. Carbon peak C1s remains the main constituent on the surface for all the samples (figure 8, table 3), but their atomic concentration decreases during natural aging. However, the trend of the

evolution of oxygen concentration is the opposite of that of carbon, in the first stage for the unaged polyethylene it was very low, after that increases during the aging process to the detriment of C1s, which points out the increasing oxidation level of the sample surfaces. For the aluminum and silicon their atomic percentage increase notably with the duration of the aging, we notice that the amount of silicon is twice as that of aluminum (table 3), this may be related to the stoichiometry of clay as illite. The percentage of the others elements present at the surface are present with very small amounts less than 1% such as nitrogen, sodium and magnesium. The exposition to natural weathering affects the state of sample surfaces, after four months the higher percentage of oxygen indicates that the surface of the material had oxidized.

Table 3. Atomic concentrations (%) of the different elements constituting the surface of the uLDPE for different exposure times under natural weathering.

			Atomic concentration (%)			
uLDPE samples	**C1s**	**O1s**	**Al2p**	**N1s**	**Si2p**	**Chemical elements (<1%)**
Non aged	99.1	0.66	0.06	0.09	0.08	-
One month	84.69	10.38	0.15	1.11	0.36	Na, Fe, Ca, S, Mg, Cl
Two months	82.54	12.45	1.09	0.28	2.13	Na, Fe, Ca, S, Mg
Three months	75.72	16.6	1.75	0.59	3.34	Na, Fe, Ca, S, Mg, Cl
Four months	64.79	23.95	3.05	0.67	5.26	Na, Fe, Ca, S, Mg, Cl

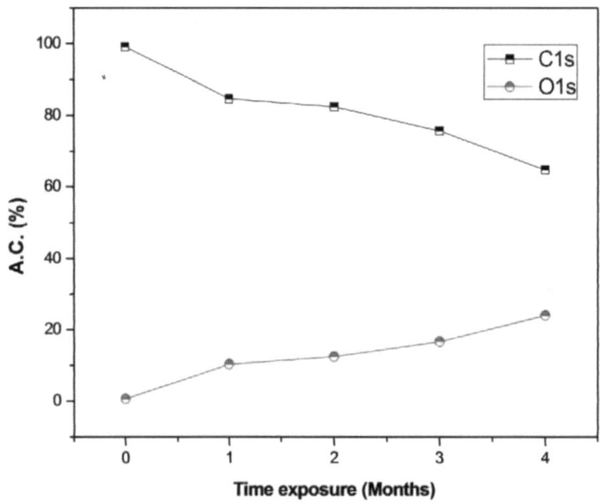

Figure 9. Evolution of percentages of the atomic concentrations of the C1s and O1s peaks according exposure time.

After the presentation of the quantitative analysis results, it is interesting to conduct a qualitative analysis, which provides more details. The study of the surface will be focused on the evolution of the main components, the peaks centered on the C1s (energy range 280-291 eV) and O1s (energy range 530-538 eV). The qualitative analysis is carried out by the decomposition of the peaks associated to C1s and O1s, thus the C1s and O1s XPS spectra will allow us to identify carbon and oxygen atoms in regard to their chemical environment, and to confirm the mechanism responsible of photo-oxidation aforesaid.

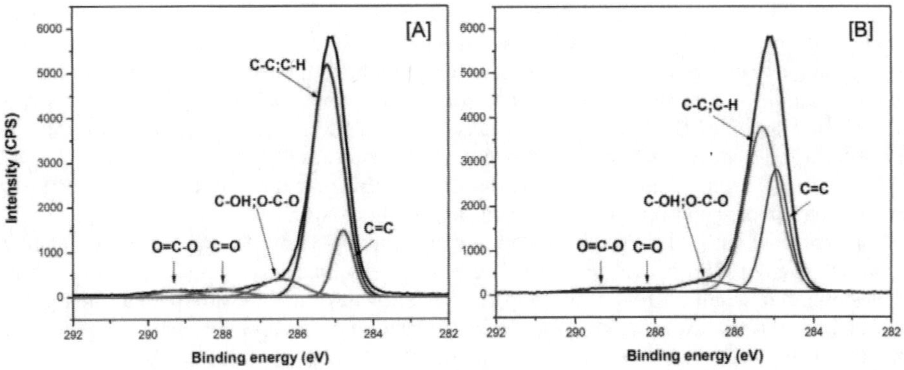

Figure 10. [A]: XPS spectra, centered on the C1s level of the uLDPE samples, aged during three months. [B]: XPS spectra, centered on the C1s level of the uLDPE samples, aged during four months.

Figure 10A shows the results of such decomposition of the peaks relative to C1s corresponding to the uLDPE sample aged during three months. The spectrum has been treated to correct the charge effect due to the insulating nature of the sample. This correction has been done by setting the -CH_n bonding energy at 285 eV. The C1s peak was deconvoluted into five subpeaks, zoom was performant in the range 291-287 eV to see clearly the subpeaks in this range (figure 10B). We distinguished five subpeaks located at 284.6, 285, 286.6, 288 and 289 eV (figure 10B).The subpeak at 285.4 eV is assigned to C-C or C-H bond, so related to the carbon atoms forming the polymer chain (Csp3) and the peak at 284.7 is a Csp2. The others subpeaks represent the bonding of carbon to oxygen, respectively C-O (286.6 eV) [58], O-C-O or C=O (288 eV) [59, 60] and O=C-O (289 eV) [60]. After four months of natural aging (figure 10B), we noticed that the subpeaks attributed to the carbon-oxygen simple bonds increased implying that surface oxidation had occurred. The peak associated to combined double and single bonding, O=C-O is more intense than the one related to double bond C=O.

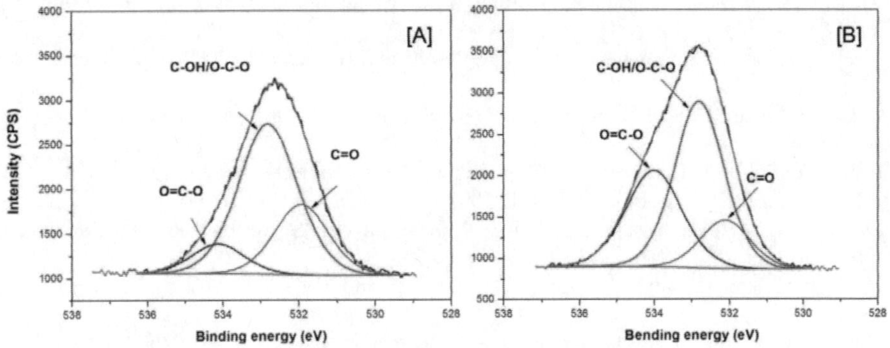

Figure 11. [A]: XPS spectra, centered on the O1s level of the uLDPE samples, aged during three months. [B]: XPS spectra, centered on the O1s level of the uLDPE samples, aged during four months.

The results of the qualitative analysis for oxygen atoms presented by the XPS spectrum of O1s (figure 11A) for the polyethylene weathered during three months can be decomposed in three subpeaks 532, 532.8 and 534 eV. The three subpeaks are related only to oxygen bonds with carbon, and are belonging respectively to C=O (532 eV) [60], C-OH, O-C-O or COOR (532.8 eV) [59,61] and O=C-O (534 eV) [62]. The others bonds like oxygen-silicon, and oxygen-aluminum are neglected. Figure 11B shows the O1s spectrum after four months of weathering, the observed differences are just quantitative; the oxygen simply bonded to the carbon atom remains the main contribution. The oxygen atom of esters groups their contribution increase to the detriment of ketones groups. The previous mechanism of photo-oxidation has been proposed for the purpose to interpret the existence of the different bonds formed by oxygen.

4. Conclusion

ATR-FTIR and XPS spectroscopy have been used to study the surface chemistry changes of the unstabilised polyethylene under natural weathering, in a quantitative and qualitative manner. By means of the X-ray photoelectron spectroscopy, the quantification of the products formed in surface of the polyethylene samples was determined with good accuracy, and then it was possible to follow the evolution of the atomic concentrations of carbon and oxygen atoms of aged polyethylene. Moreover, various chemical elements present on the polymer surface with low concentrations were unveiled. Through the deconvolution of the C1s and O1s peaks, the qualitative analysis provides information about the chemical environment of the carbon and oxygen atoms, and then we come to know the nature of bonds established. Comparing these results to that of ATR-FTIR, we conclude the mechanism of photo-oxidation of our samples is made according to Norrish II.
ATR-FTIR spectroscopy has been used to understand structural modifications, and to monitor carbonyl and vinyl groups formations, and to determine crystallinity changes during weathering. Structural changes after weathering were investigated by following carbonyl group, and vinyl group formation, which increase as exposure increase. The calculated vinyl group index can also be used to indicate polymer chain scission. Indeed, chain scission occurs during exposure, as evidence by growth of carbonyl and vinyl groups. Furthermore, an increase in crystallinity indicates the polyethylene chain scission during the photo-oxidation. Increase in crystallinity can occur if chain scission occurs; consequently, the polymer experienced a secondary crystallization, which takes place in the amorphous phase. The origin of the chain scission mainly occurs through photolysis of ketones according Norrish II mechanism.

Acknowledgments

Special thanks to the Innovation city of USMBA (Morocco), Nanosciences African Network, University of South Africa Department of Physics, iThemba Labs (South Africa), African Laser Center and Africa Graphene Center.

References

[1] Feldman D 2002 *J. Poly. Environ.* **10** 163
[2] Rabek JF 1996 P*hys. charact. applications*
[3] Scoponi M, Cimmino S and Kaci M 2000 *Polymer.* **41** 7969
[4] Hassini N, Guenachi K, Hamou A, Saiter J M, Marais S and Beucher E 2002 *Poly. Degrad. Stab.* **75** 247
[5] Szép A, Anna P, Csontos I, Marosi Gy, Matko Sz and Bertalan Gy 2004 *Poly. Degrad. Stab.* **85** 1023
[6] Sqalli Houssaini D, Rjeb A, Rjeb M, Sayouri S, Roy D, Massey S and Sadoun T 2007 *J. Catal. Mat. Env.* **VI** 213
[7] Boulos Y, Dehbi A, Hamou A and Marc Saiter J 2008 *Mater Des.* **29** 2017
[8] Massey S, Adnot A, Rjeb A and Roy D 2008 *Plast. Rubber Compos.* **37**:7
[9] Darhouri M, Sqalli Houssaini D, Rjeb A, Rjeb M, Sayouri S, Roy D, Massey S, Sadoun T and Laoutid F 2009 *Phys. Chem. News.* **47** 40
[10] Hamzah M, Rjeb A, Sqalli Houssaini D, Sayouri S and Massey S 2011 *Phys. Chem. News.* **58** 73
[11] Hamzah M, Rjeb A, Sqalli Houssaini D, Sayouri S and Darhouri M 2013 *Phys. Chem. News.* **70** 7
[12] Carrasco F, pages P, Pascual S and Colom X 2001 *Eur. Poly. J.* **37** 1457
[13] Kupper L, Gulmine JV, Janissek PR and Heise HM 2004 *Vib. Spectrosc.* **34** 63
[14] Dehbi A, Bouaza A, Hamou A, Boulos Y and Marc Saiter J 2010 *Mater. Des.* **31** 864
[15] Grabmayer K, Wallner GM, Beißmann S, Braun U, Steffen R, Nitsche D, Roder B, Buchberger W and Lang RW 2014 *Poly. Degrad. Stab.***109** *40*
[16] Schwarzinger C, Hintersteiner I, Schwarzinger B, Buchberger W and Moser B 2015 *J. Anal. Appl. Pyrolys..* **113** 315
[17] Jabarin SA and Lofgren EA 1994 *J. Appl. Poly. Sci.* **53** 411
[18] Charlesby A and Partridge RH 1965 *Proceedings of the Royal Society of London. Series A: Math. Phys. Sci.***283** 312

[19] Allen NS, McKellar JF and Phillips GO 1974 *Poly. Lett. Edition* **12** 253
[20] Peacock AJ 2000 Handbook of polyethylene Structures, Properties, and Applications. By Marcel Dekker
[21] Chabira SF, Sebaa M, Huchon R and De Jeso B 2006 *Poly. Degrad. Stab.* **91** 1887
[22] Wypych G 1995 Handbook of material weathering.2nd ed. Ontario, Canada: Chem Tec Publishing
[23] Zenkiewicz M, Rauchfleisz M and Czuprynska J 2003 *Rad. Phys. Chem.* **68** 799
[24] Tertinnikov ON, Fujita SI, Ogata S and Ikada Y 1999 *J. Polym. Sci. Part B: Polym. Phys.* **37** 1503
[25] Zerbi G, Gallino G, Del Fanti N and Baini L 1989 *Polymer.* **30** 2324
[26] Gulmine JV, Janissek PR, Heise HM and Akcelrud L 2002 *Poly. Test.* **21** 557
[27] Stuart B 2004 *Chapter 4 Organic Molecules*. Wiley **71**
[28] Shi LS, Wang LY and Wang YN 2006 *Eur. Poly J.* **42** 1625
[29] Wu H and Fan J 2008 *Poly. Test.* **27** 122
[30] Fu Y and Lim LT 2012 *Poly. Test.* **31** 56
[31] Oldak D, Kaczmarek H, Buffeteau T and Sourisseau C 2005 *J. Mater. Sci.* **40** 4189
[32] Feuilloley P, César G, Benguigui L, Grohens Y, Pillin I, Bewa H, Lefaux S and Jamal M 2005 *J. Poly. Env.* **13** 349
[33] Stark N M and Matuana L M 2004 *Poly. Degrad. Stab.* **86** 1
[34] Gulmine JV, Janissek PR, Heise HM and Akcelrud L 2003 *Poly. Degrad. Stab.* **79** 385
[35] Craig IH, White JR, Shyichuk AV and Syrotynska I 2005 *Poly. Eng. Sci.* **45** 579
[36] Vasile C and Pascu M 2005 *Practical guide to polyethylene*. Rapra Technology
[37] Tidjani A 2000 *Poly. Degrad. Stab.* **68** 465
[38] David C, Trojan M, Daro A and Demarteau W 1992 *Poly. Degrad. Stab.* **37** 233
[39] Khabbaz F, Albertsson AC and Karlsson S 1998 *Poly. Degrad. Stab.* **61** 329
[40] Yang R, Li Y and Yu J 2005 *Poly. Degrad. Stab.* **88** 168
[41] Guadagno L, Naddeo C, Vittoria V, Camino G and Cagnani G 2001 *Poly. Degrad. Stab.* **72** 175
[42] Allen N S, Edge M, Holdsworth D, Rahman A, Catalina F, Fontan E, Escalona A M and Sibon F F 2000 *Poly. Degrad. Stab.* **67** 57
[43] Valadez-Gonzalez A, Cervantes-Uc JM and Veleva L 1999 *Poly. Degrad. Stab.* **63** 253
[44] Douminge L, Mallarino S, Cohendoz S, Xavier F and Bernard J 2010 *Curr. Appl. Phys.* **10** 1211
[45] Kaci M, Sadoun T and Cimmino S 2001 *Int J Polym Anal. Charact.* **6** 455
[46] Mendes LC, Rufino ES, De Paula FOC and Torres Jr AC 2003 *Poly. Degrad. Stab.* **79** 371
[47] Grossman DM 1998 *Surf. Coat. Aus.* **25** 6
[48] Allen NS and McKellar JF 1975 *Chem. Soc. Rev.* **4** 533
[49] Gijsman P and Dozeman A 1996 *Polym. Degrad. Stab.* **53** 45
[50] Searle N D 2000 Handbook of Polymer Degradation chapter 16. Dekker, 605-643
[51] Charlesby A and Partridge R H 1965 *Series A: Math. Phys. Sci.* **283** 312
[52] Carlsson D J and Wiles D M 1969 *Macromolecules.* **2** 587
[53] Pagès P, Carrasco F, Surina J and Colom X 1996 *J. Appl. Polym. Sci.* **60** 153
[54] Iring M, Tüdos F, Fodor Z and Kelen T 1980 *Polym. Degrad. Stab.* **2** 143
[55] Costa L, Luda MP and Trossarelli L 1997 *Polym. Degrad. Stab.* **58** 41
[56] Khelidj N, Colin X, Audouin L, Verdu J, Monchy-Leroy C and Prunier V 2006 *Polym. Degrad. Stab.* **91** 1598
[57] El Ouahabi M, Daoudi L and Fagel N 2014 *Clay. Minerals.* **49** 35
[58] Rats D, Vandenbulcke L, Herbin R, Benoit R, Erre R, Serin V and Sevely J 1995 *Thin Solid Films* **270** 177
[59] Brown N M D, Hewitt JA and Meenan B J 1992 *Surf. Inter. Anal.* **18** 87
[60] Rjeb M, Labzour A, Rjeb A, Sayouri S, Chafik El Idrissi M, Massey S, Adnote A and Roy D 2004 J. *Phys.* Condens. Matter. **5** 168
[61] Rjeb A, Letarte S , Tajounte L , Chafik El Idrissi M ,Adnot A , Roy D, Claire Y and Kaloustian J 2000 J. Electron. Spectrosc. Relat. Phenom. **107** 221
[62] Poleunis C, Weng LT, Sclavons M, Bertrand P, Franquinet P, Legras R and Carlier V 1995 J. Adhes. Sci. Technol. **9** 859

The effect of strontium and silver on the bioactivity of a quaternary bioglass in the system SiO_2-CaO-Na_2O-P_2O_5

S Herradi[1,*], S Bouhazma[1], S Chajri[1], M Khaldi[1], A El Hachadi[2], B El Bali[1] and M Lachkar[1]

[1]Engineering Laboratory of Organometallic and Molecular Materials, Faculty of Sciences, University Sidi Mohamed Ben Abdellah, Po. Box 1796 (Atlas), 30000 Fez, Morocco
[2]Laboratory of Process, Signals, Industrial Systems, Computing, High School of Technology, University Cadi Ayyad, Po. Box 549, Safi, Morocco

[*] Email: Smaiel_her@hotmail.fr

Abstract. Bioactive glasses with $46SiO_2$-28CaO-$20Na_2O$-$6P_2O_5$ (46S0), $46SiO_2$-28CaO-$14Na_2O$-2SrO-$4Ag_2O$-$6P_2O_5$ (46S2), $46SiO_2$-28CaO-$14Na_2O$-4SrO-$2Ag_2O$-$6P_2O_5$ (46S4) and $46SiO_2$-28CaO-$14Na_2O$-6SrO-$6P_2O_5$ (46S6) compositions were prepared by the sol-gel method. The obtained bioglasses were sintered at 600°C for 4 hours. The structural and bioactive properties of these sintered glasses have been investigated by using X-ray diffraction, Fourier Transform Infrared Spectroscopy (FTIR). In vitro bioactivity was investigated by soaking the bioglasses in SBF for 14 and 21 days at 37°C. XRD and FTIR have shown that only the 46S0 and 46S6 bioglasses were bioactive by the formation of hydroxycarbonated apatite layer (HCA) and that bioactivity of 46S6 is better than of 46S0, however, no bioactivity have been noticed for 46S2 and 46S4 bioglasses and only AgCl has been formed after soaking in SBF as evidenced by XRD patterns. The measured wettability using a static sessile drop technique has demonstrated a high hydrophilicity for both 46S0 and 46S6 bioglasses.

1. Introduction

Bioactive glass is a material that degrades both in physiological environment and in simulated body fluid (SBF) leading to the formation of a HCA layer which act as a bonding surface between the host tissue and the bioactive glass [1-3]. The first bioactive glass (Bioglass® 45S5) was reported by Hench *et al* in 1971, it is a melt-derived bioactive glass composed of 45wt% SiO_2, 24.5wt% CaO, 24.5wt% Na_2O and 6wt% P_2O_5 [4, 5]. Since then, silica-based glasses have been used as hard and soft tissue treatment biomaterials due to their biocompatibility [6]. Sol-gel processing has been intensely used as an alternative method for the synthesis of new bioactive and bioresorbable glasses owing to many advantages such as homogeneity, purity, low temperature synthesis and better control over composition and microstructure [7-10]. Particularly, sol gel processing has enabled the expanding of the range of SiO_2 content in which bioactivity could be observed, from 60 mol% in melt-derived Bioglass® to 90 mol% in sol-gel derived glass. Moreover, it has been reported that CaO-P_2O_5-SiO_2 glass produced by the sol-gel technique is more bioactive than the melt-derived glass of the same composition [10].

Bioactive glasses in the system SiO_2–CaO–P_2O_5–Na_2O have shown higher bioactivity along with promoting bone formation by changing the intracellular ionic concentrations thus mediating cell metabolism [11-13]. Recently, the trend has been shifted toward the preparation of smart bioglasses with additional features such as anti-inflammatory and bone growth ability [14]. In this regard, inclusion of silver into the bioglass composition will be beneficial owing to its antibacterial properties [15]. In fact, silver foils have been used for surgical wound and broken bones treatment [16]. Nowadays, silver-substituted bioactive glasses are used as a coating for medical suture which play a dual role in term of bioactivity and antibacterial properties [17]. In the other hand, the beneficial effects of strontium on bone cells and bone formation in vivo are well documented in the literature [18, 19]. Oral administration of strontium has also been shown to be promising for treating osteoporosis by stimulating osteoblasts to form new bone and hindering bone resorbing by osteoclasts [20, 21]. Therefore, a glass that contain both strontium and silver in its composition may combine bone bonding, osteoblast stimulating and anti-inflammatory properties.

Herein we report the synthesis of bioglasses on a mole percentage basis in the systems $46SiO_2$-$28CaO$-$20Na_2O$-$6P_2O_5$ (46S0), $46SiO_2$-$28CaO$-$14Na_2O$-$2SrO$-$4Ag_2O$-$6P_2O_5$ (46S2), $46SiO_2$-$28CaO$-$14Na_2O$-$4SrO$-$2Ag_2O$-$6P_2O_5$ (46S4) and $46SiO_2$-$28CaO$-$14Na_2O$-$6SrO$-$6P_2O_5$ (46S6). We then investigated their bioactivity by soaking them in SBF solution. The bioactive glasses were then subjected to angle contact and open porosity measurement.

2. Materials and methods

2.1. *Materials*

The following chemicals were used as precursors for the synthesis of the sol-gel glasses, i.e. 46S0, 46S2, 46S4 and 46S6: tetraethylorthosilicate (TEOS) (Aldrich, 99%), triethylphosphate (TEP) (Eastman, 99.8%), sodium nitrate (Sigma-Aldrich, 99%), strontium nitrate (SIGMA-ALDRICH, ≥ 98%), calcium nitrate tetrahydrate (LobaChemie, 98%) and silver nitrate (VWR Chemicals, ≥99.5%).

2.2. *Preparation of bioglasses*

The exact nominal compositions for the four bioglasses are given in table 1. In brief, the molar ratios of TEOS, TEP, $NaNO_3$, $Sr(NO_3)_2$, $AgNO_3$ and $Ca(NO_3)_2.4H_2O$ were designed according to the molar ratio of SiO_2, P_2O_5, CaO, SrO, Na_2O and Ag_2O in 46S0, 46S2, 46S4 and 46S6 bioglasses. To achieve a clear sol, we used 2N of HNO_3 to catalyze the TEOS and TEP hydrolysis, using a molar ratio of $H_2O/(TEOS+TEP)=8$. TEOS, deionized water, and nitric acid, were successively mixed and the mixture was allowed to react for 1h under continuous stirring. Then appropriate amount of each of the remaining reagents was added and allowed to react completely for 30 min in the following sequence: (TEP), $Ca(NO_3)_2.4H_2O$, $NaNO_3$, $AgNO_3$ and/or $Sr(NO_3)_2$ (for the 46S0 glass neither $AgNO_3$ nor $Sr(NO_3)_2$ were added). After the addition of the last reagent, the mixture was stirred for one additional hour to ensure homogeneity. The solutions were then introduced in a hermetically sealed container, where they were allowed to gel at 25°C for 3 days and then were aged at 80°C for 3 days. Then, the aged gels were dried at 150°C for 52 hours (gelation, aging and drying of samples were conducted in a programmable oven). The obtained dry gels were ground using an agate mortar. Subsequently, the stabilization of the powders was carried out in an electric furnace at 600°C for 4 hours. Finally, these powders were ground again to break down the agglomerates and to enhance the reactivity.

Table 1. Nominal composition (mol %) of the synthesized bioglasses.

Sample	SiO_2	CaO	P_2O_5	Na_2O	Ag_2O	SrO
46S0	46	28	6	20	0	0
46S2	46	28	6	14	4	2
46S4	46	28	6	14	2	4
46S6	46	28	6	14	0	6

2.3. *In vitro bioactivity assessment of bioglasses*

The bioactivity of the prepared bioglasses was investigated by in vitro test according to the method described by Kokubo and Takadama [22]. The test was performed by immersing 0.1g of each samples in 45 mL of SBF solution (pH=7.4) contained in a polyethylene container and incubated at 37°C in a static condition for time periods of 14 and 21 days. A Comparison between ionic concentrations of SBF and human blood plasma (mmol/L) is shown in table 2 [22]. After soaking, the samples were filtered and dried in air.

Table 2. Comparison of ionic concentrations of SBF and human blood plasma (mmol/L).

Ion	Simulated Body Fluid	Blood plasma
Na^+	142.0	142.0
K^+	5.0	5.0
Mg^{2+}	1.5	1.5
Ca^{2+}	2.5	2.5
Cl^-	148.8	103.0
HCO_3^-	4.2	27.0
HPO_4^{2-}	1.0	1.0
SO_4^{2-}	0.5	0.5

2.4. *pH behavior*

The pH behavior of the SBF solution, immediately after removing the powder from it, was investigated. The used instrument for pH measurement was calibrated each time with standard buffer solutions of pH 4, 7 and 9.23.

2.5. *Characterization of the bioglasses*

The powders, before and after immersion in SBF, were analysed by X-ray diffraction using Discover model equipped with a monochromatized Cu-Kα radiation (λ = 1.5418 Å). The diffraction patterns were obtained in the 2θ range 20°-80°. Fourier transform infrared spectroscopy (FTIR: Bruker VERTEX 70 spectrometer) was used to determinate the functional groups of the obtained powders before and after soaking in SBF. The spectra were recorded from 400 to 4000 cm^{-1} at 4 cm^{-1} resolution.

2.6. *Contact angle measurements*

Contact angle measurements were performed on 46S0 and 46S6 bioglasses to investigate the degree of wettability. To do so, two pellets of each glass were prepared by pressing 0.2 g of each sample in a Specac hydraulic press at 8 tms for 60s. The wettability was determined by the sessile-drop technique using a goniometer (GBX-France). It's a static contact angle measurement method, which consists in putting down a liquid drop on the solid plate we want to characterize its surface by measuring the contact angle made by the drop on this surface. As it is shown in figure 1, when a drop of a liquid is putted down on a solid surface; three phases system occurs: solid, liquid and gas [23]. The used liquids were deionized water and SBF solution. A drop (~ 1μL) of water or SBF is placed on the glass pellet, and the image is immediately taken after 5s by the camera witch is coupled with a computer. The measurements were repeated by depositing at least three drops on each pellet and the results were given by the average.

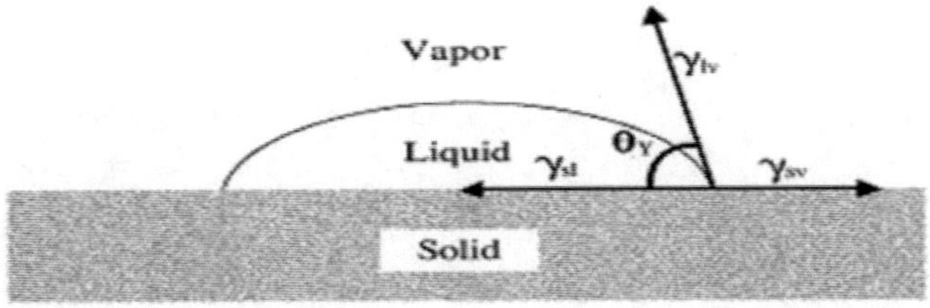

Figure 1. Static contact angle measurement with the sessile drop method where γ_{sv}, γ_{sl} and γ_{lv} represent the "surface tensions" of the interface solid/gas, solid/liquid and liquid/gas, respectively, and θ represents the contact angle [23].

2.7. Open porosity measurements

The open porosity was calculated using Arthur method (equation (1)).

$$P_O = \frac{(m2-m1)}{(m2-m3)} * \frac{\rho watter}{\rho xyléne} \quad (1)$$

Where m_1 represent the weight of sample in air, m_2 and m_3 are the weights after soaking the sample in xylene and water respectively, ρ_{water} is the density of water and ρ_{xylene} is the density of xylene.

3. Results and discussion

3.1. XRD patterns

The XRD patterns of the glasses 46S0, 46S2, 46S4 and 46S6, before and after soaking in SBF for 14 and 21 days, are provided in figure 2, figure 3, figure 4 and figure 5 respectively. For 46S0 glass, as it can be seen in figure 2, before soaking in SBF the powder was composed solely of a well crystallized $Na_2Ca_2Si_3O_9$ phase, however, after 14 days in SBF almost all peaks of this phase have almost disappeared and replaced by those corresponding to HA phase. Further, after 30 days in SBF the remaining peaks of $Na_2Ca_2Si_3O_9$ phase has disappeared, only a very weak peak exist at 20.19° witch indicate that almost all the powder has converted to HA and that $Na_2Ca_2Si_3O_9$ phase barely exist [24-28]. The crystallization did not inhibit the bioactivity of the 46S0 glass since the $Na_2Ca_2Si_3O_9$ phase has been reported to be biodegradable [24].

Interestingly, the 46S2 and 46S4 glasses show another behavior either before or after soaking in SBF. In fact, as it is depicted in figure 3 and figure 4, the as-prepared glasses patterns indicate that they are composed of metallic Ag, $Na_2Ca_2Si_3O_9$ and $CaSiO_3$ [24, 28-30], whereas, after both 14 and 21days, we have noticed the formation of AgCl phase rather than HA. The formation of AgCl on silver substituted bioglasses surface was reported elsewhere [31, 32]. This compound forms when Ag^+ ions in the glass network reacted with Cl^- ions in SBF solution.

The 46S6 bioglass shows the higher bioactivity as it could be seen in figure 5. Unlike the 46S0 bioglass, the as-prepared 46S6 bioglass is composed of a weakly crystallized $Na_2Ca_2Si_3O_9$ phase which would facilitate its degradation in SBF solution. In fact, after 14 and 21days in SBF the 46S6 powder patterns show only peaks corresponding to HA phase.

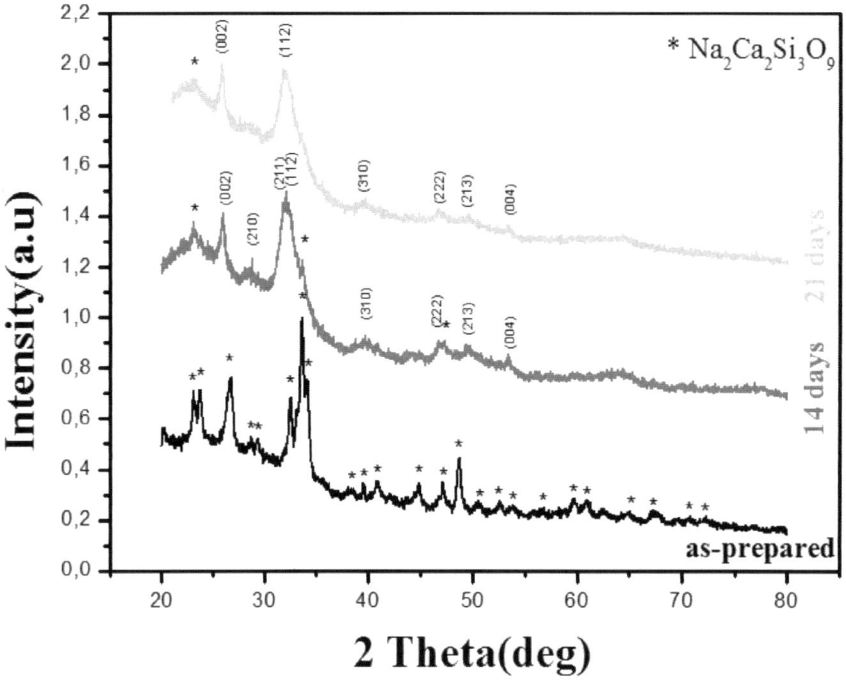

Figure 2. XRD patterns of the 46S0 bioglass before and after 14, and 21 days soaking in SBF.

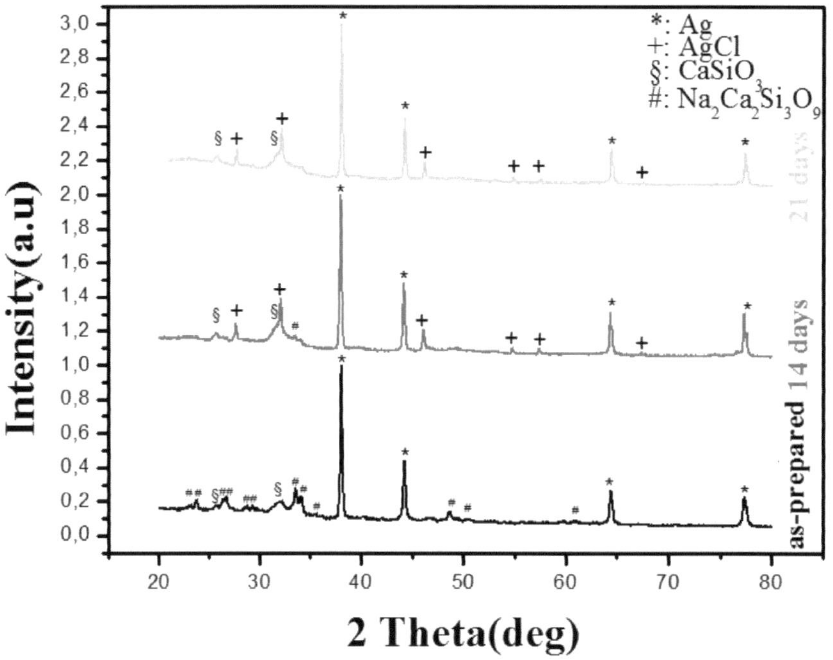

Figure 3. XRD patterns of the 46S2 bioglass before and after 14, and 21 days soaking in SBF.

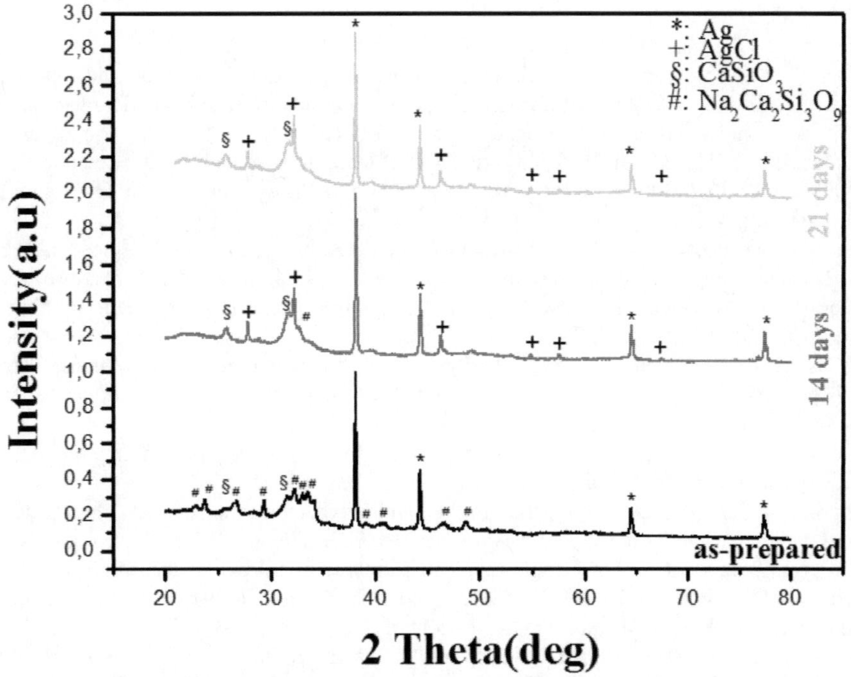

Figure 4. XRD patterns of the 46S4 bioglass before and after 14, and 21 days soaking in SBF.

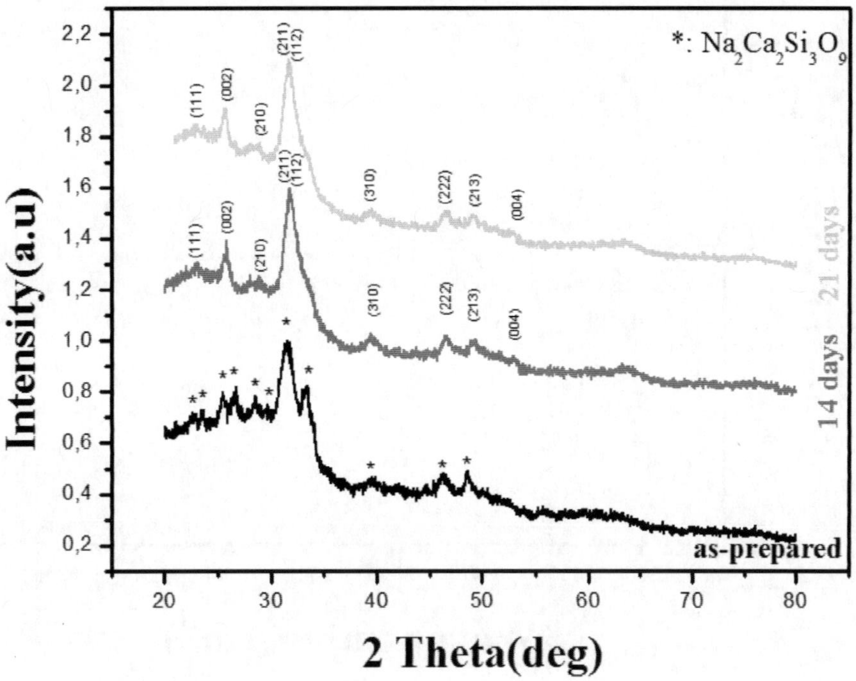

Figure 5. XRD patterns of the 46S6 bioglass before and after 14, and 21 days soaking in SBF.

3.2. FTIR spectra

The figure 6 shows the spectra of the as prepared 46S0 and 46S6 samples. For both glass powders, the band located in 1016 cm^{-1} correspond to Si–O–Si asymmetric stretching vibration, the band located at 914 cm^{-1} is related to the stretching of non-bridging oxygen atoms in Si-O bond. The band at 443 and 446 cm^{-1} for 46S0 and 46S6 samples respectively are attributed to Si–O–Si symmetric bending mode. The bands located at 518, 575 and 614 cm^{-1} are due P-O bending and the band located at 695 and 753 cm^{-1} in 46S0 and 46S6 spectrum respectively are assigned to the symmetrical Si–O–Si stretching in $Na_2Ca_2Si_3O_9$ phase [24, 33, 34].

Interestingly, the spectra of 46S0 glass shows well resolved bands compared to those of 46S6 glass and this is because the $Na_2Ca_2Si_3O_9$ phase is more crystallized in 46S0 glass as is it confirmed by the XRD patterns which explains the reason why this crystalline phase has degraded completely after putting the 46S6 glass 14 days in SBF.

The IR spectra of the bioglasses 46S0 and 46S6 after soaking for 14 and 21 days in SBF are shown in figure 7 and figure 8 respectively. For 46S0 glass soaked for 14 days, new bands appear at 566, 606, 1085, 1443 and 875 cm^{-1} and are due to P–O antisymmetric bending, ν_4 bending mode in PO_4^{3-}, P–O antisymmetric stretching, the absorption of carbonate group (ν_3) and ν_2 bending mode in CO_3^{2-} respectively which are due to the formation of a new HCA layer [35, 36].

The band located at 1639 cm^{-1} is ascribed to the bending vibration mode in adsorbed water. As the time of soaking increase to 21 days, new bands appears at 793, 875 and 955 cm^{-1} ascribed respectively to Si–O–Si symmetric stretching of bridging oxygen atoms between two adjacent SiO_4 tetrahedra and ν1 stretching mode, in PO_4^{3-} in HCA [35, 36]. Interestingly, the band at 793 cm^{-1} which belongs to Si-O-Si symmetric stretching, indicates the formation of silica-rich layer [37]. The same bands are noticed for 46S6 glass spectra except the fact that the spectrum for 14 days is more resolved and shows clearly the bands at 955 cm^{-1} attributed to ν_1 stretching mode, in PO_4^{3-} in HCA. These results correlate with the XRD patterns and confirm that the 46S6 bioglass is more bioactive than 46S0.

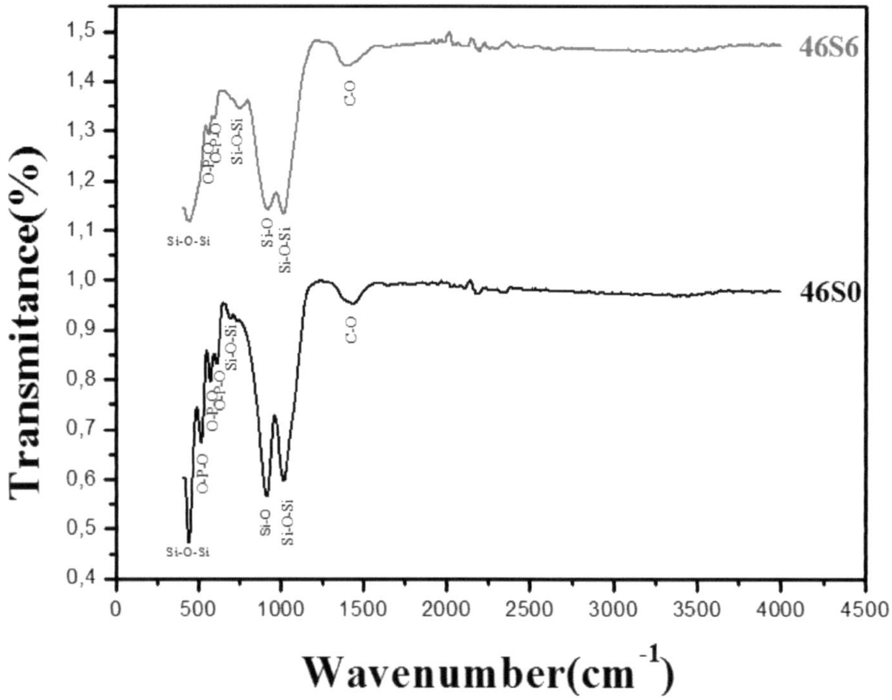

Figure 6. FTIR spectra of the 46S0 and 46S6 bioglasses before soaking in SBF.

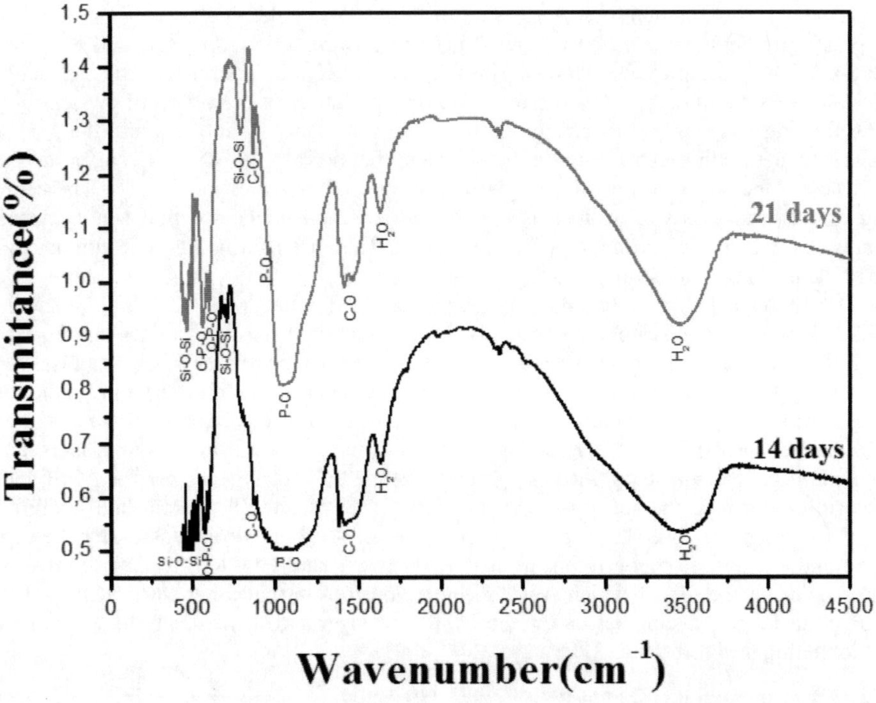

Figure 7. FTIR spectra of 46S0 bioglass after 14 and 21days soaking in SBF.

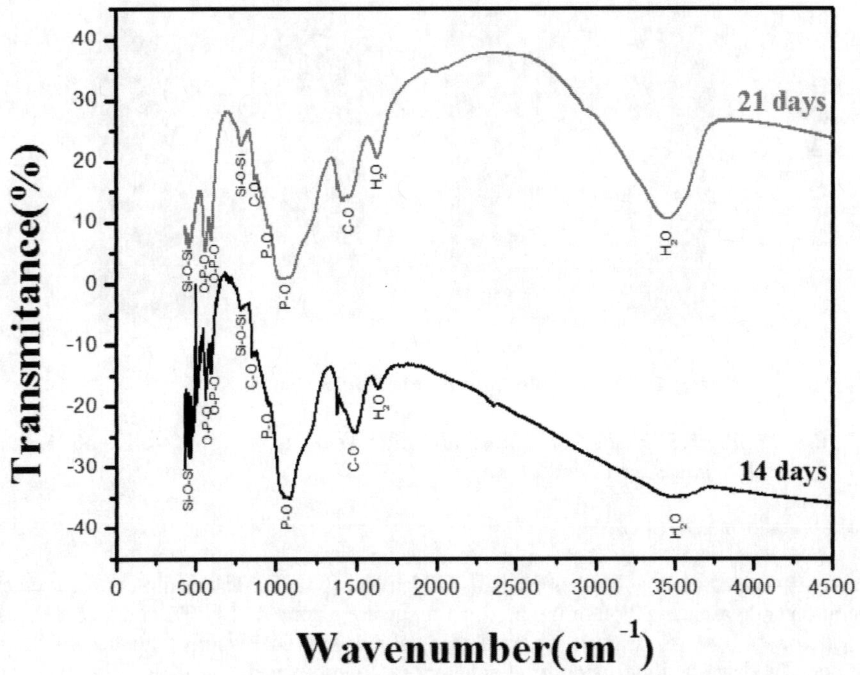

Figure 8. FTIR spectra of 46S6 bioglass after 14 and 21days soaking in SBF.

3.3. *Changes in pH*

The pH change in SBF solution after immersion of the glass samples is shown in figure 9. After 14 days of soaking in SBF solution, the measured pH values were 8.95, 8.68, 8.82 and 8.73 for 46S0, 46S2, 46S4 and 46S6 samples respectively. This increase in pH was first reported by Hench *et al.* [1] as the first stage in the formation of an apatite-like layer on bioactive melt-derived glasses containing $CaO\text{-}SiO_2$. During this stage , interchange between the Ca^{2+} ions of the glass and the H_3O^+ of the solution took place leading to an increase in pH. Then, the increase in hydroxyl concentration of the solution leads to the attack in the silicate glass network and formation of silanols. The continuous releasing of Ca^{2+} ions leads to an increase in its concentration in the solution hence enabling the precipitation of HCA nuclei on the top of silanols layer. The nuclei formed grew continuously at the expense of the ions in the solution.

After 21 days of soaking in SBF solution, the measured pH values were 9.33, 9, 9.06 and 9.02 for 46S0, 46S2, 46S4 and 46S6 samples respectively. This second increase in pH values may be explained by the fact that the silanols layer is permeable which would fasten the release of Ca^{2+} and Na^+ ions into the surrounding solution thus leading to a fast increase in the concentration of OH^- ions in the solution. The release of ions trough the silanols layer would continue until its depletion from the glass network. Furthermore, the increase in pH after 21 days of soaking in SBF confirm that our glasses would be more and more biodegradable by increasing soaking time. The change in pH for 46S0 and 46S6 glasses corroborates with the results shown by XRD patterns and IR spectra and confirms their bioactivity. However, the observed increase in pH values for 46S2 and 46S4 glasses may be explained by only a rapid interchange between ions in the glass network and H_3O^+ ions leading to the formation of silanols layer on the top of which AgCl precipitation took place rather than HCA. It has been demonstrated that AgCl precipitates on the top of silanols layer and slows down the glass dissolution rate thus inhibiting the bioactivity [31].

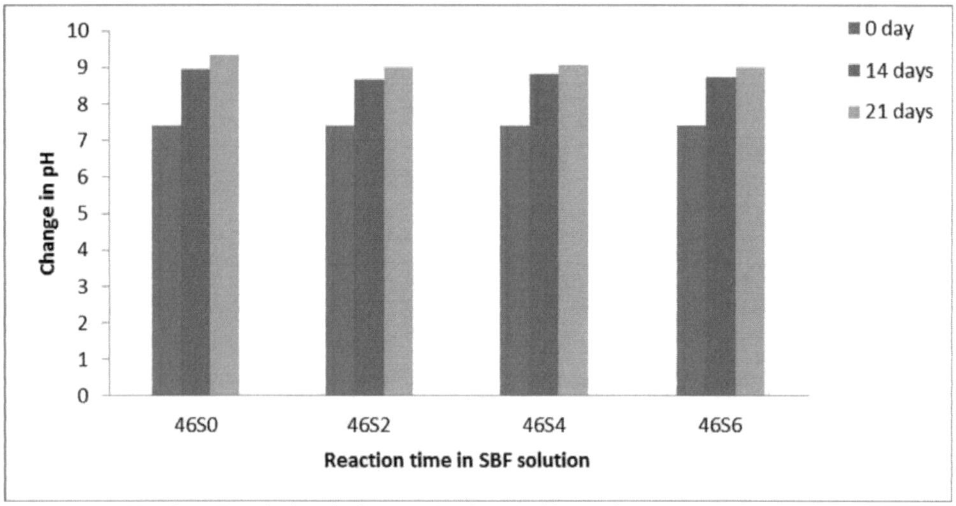

Figure 9. pH of SBF solution before and after soaking of 46S0, 46S2, 46S4 and 46S6 bioglasses for 14 and 21 days.

3.4. *Open porosity measurements*

An ideal biomaterial should possess an interconnected porous structure for cell seeding, tissue ingrowth and vascularization, as well as for nutrient delivery [38, 39]. The minimum requirements for cell penetration and vascularization of the ingrown tissue are a porosity of 90% and a pore diameter of at least 100 μm [40, 41]. It is well reported that sol-gel derived bioactive glasses show a better bioactivity and biodegradability owing to its inherent mesoporous texture [42-44]. In this regard, we have measured the open porosity of 46S0 and 46S6 bioactive glasses, the values found are shown in table 3. Those porosity values are mainly due to the decomposition of nitrates and the removal of solvents trapped within the dried gel during the stabilization step [45, 46].

Moreover, we observe that the porosity is almost the same for both samples, this is due to the chemical similarity of strontium and calcium ions, such as ionic radius (Ca: 0.94 Å; Sr: 1.16 Å). Xiang *et al* reported that SrO/CaO substitution has little effect on the glass network structure and that the increase in the dissolution rate of strontium substituted glasses was mainly due to increase of both molar volume and density [14].

Table 3. Measured open porosity of 46S0 and 46S6 bioglasses.

Sample	**46S0**	**46S6**
Open porosity (%)	60.52	60.88

3.5. Contact angle measurements

In general, the surface properties of a solid determine its behavior towards gases, fluids and other solids [43]. In the case of a biomaterial, wettability is an important property that affects fibroblast and endothelial cell attachment as well as protein adsorption [47-49]. In particular, hydrophilic biomaterials are suitable for cells attachment while hydrophobic biomaterials improve proteins adsorption [50]. Vogler has reported that the material hydrophilic/hydrophobic character can be determined by measuring its contact angle [51]. He stated that when contact angle is above 65°, the material is hydrophobic otherwise, it is hydrophilic. The contact angles of the 46S0 and 46S6 bioglasses, using water or SBF, are shown in table 4. As is it clear, both samples present very low contact angles either using water or SBF, however, the measured contact angle for 46S0 sample using water was lower than for 46S6. Inversely, using SBF solution the measured contact angle for 46S6 sample was lower than for 46S0 sample. These lower contact angles indicate that the 46S0 and 46S6 bioglasses possess superhydrophylic character; this is related to the high release of Na^+ ions from the glass network [52].

Table 4. Contact angles of 46S0 and 46S6 bioglasses.

Samples	**Contact angle (°)**	
	Water	**SBF**
46S0	6.7 ± 1.45	14.1 ± 0.31
46S6	13.3 ± 0.41	7.6 ± 0.92

4. **Conclusion**

In this work, bioglass powders in SiO_2-CaO-Na_2O-P_2O_5, SiO_2-CaO- Na_2O-Ag_2O-SrO-P_2O_5 and SiO_2-Na_2O-CaO-SrO-P_2O_5 systems were obtained by a sol-gel technique. The X-ray diffraction patterns have shown that only 46S0 and 46S6 were bioactive and that the 46S6 glass is more bioactive than the 46S0 glass, whereas, the 46S2 and 46S4 bioglasses aren't bioactive due to the precipitation of AgCl compound rather than HCA. Furthermore, the IR spectra have confirmed the enhancement of bioactivity due to the inclusion of Sr^{2+} ions into the SiO_2-CaO-Na_2O-P_2O_5 system. The porosity was slightly increased by strontium incorporation owing to the similarity in radius between Sr^{2+} and Ca^{2+} ions. Moreover, the measured contact angles for the 46S0 and 46S6 bioglasses have demonstrated their superhydrophilic character, which would make them very useful for bone tissue engineering scaffolds.

Acknowledgements- The authors would like to acknowledge the support and technical assistance of Interface Regional University Center (University Sidi Mohammed Ben Abdellah, Fez, Morocco).

References

[1] Hench LL and West JK 1996 *Biological applications of bioactive glasses* Life Chemistry Reports 13 (New York: Harwood Academic) pp 187–241
[2] L.L. Hench 1991 *J. Am. Ceram. Soc.* **74** 1487–510
[3] Dietrich E, Oudadesse H, Lucas-Girot A and Mami M 2009 *J. Biomed. Mater. Res.* A **88** 1087–96
[4] Hench L L, Splinter R J, Allen W C and Greenlee T K 1971 *J. Biomed. Mater. Res.* **5** 117–41
[5] Shirtliff V J and Hench L L 2003 *J. Mater. Sci.* **38** 4697–707
[6] Hench L L and Wilson J 1984 *Science* **226** 630–6
[7] Hench L L and West J K 1990 *Chem. Rev.* **90** 33–72
[8] Hench L L, Wheeler D L and Greenspan D C 1998 J. Sol-Gel Sci. Technol. **13** 245–50
[9] Brinker C J and Scherer G W 1990 *Sol–Gel Science: The Physics and Chemistry of Sol–Gel Processing* (New York: Academic Press) p 11
[10] Li R, Clark A E and Hench L L 1991 *J. Appl. Biomater.* **2** 231–9
[11] Hoppe A, Guldal N S and Boccaccini A R 2011 *Biomaterials* **32** 2757–74
[12] Hench L L 1998 *J. Am. Ceram. Soc.* **81** 1705–28
[13] Saravanapavan P, Jones J R, Pryce R S and Hench L L 2003 *J. Biomed. Mater. Res.* **66** 110–9
[14] Xiang Y and Du J 2011 *Chem. Mater.* **23** 2703–17
[15] Catauro M, Raucci M G, De Gaetano F and Marotta A 2004 *J. Mater. Sci. - Mater. Med.* **15** 831-7
[16] Balamurugan A, Balossier G, Laurent-Maquin D, Pina S, Rebelo A H S, Faure J and Ferreira J M F 2008 *Dent. Mater.* **24** 1343-51
[17] Blaker J J, Nazhat S N and Boccaccini A R 2004 *Biomaterials*, **25** 1319–29
[18] Marie P J, Ammann P, Boivin G and Rey C 2001 *Calcif. Tissue Int.* **69** 121–9
[19] Marie P J 2006 *Bone* **38** 10–4
[20] Meunier P J, Slosman D O, Delmas P D, Sebert J L, Brandi M L, Albanese C, Lorenc R, Pors-Nielsen S, De Vernejoul M C, Roces A and Reginster J Y 2002 *J. Clin. Endocrinol. Metab.* **87** 2060–6
[21] Bonnelye E, Chabadel A, Saltel F and Jurdic P 2008 *Bone* **42** 129–38
[22] Kokubo T and Takadama H 2006 *Biomaterials* **27** 2907–15
[23] Lotfi M, Nejib M and Naceur M 2013 Cell Adhesion to Biomaterials: Concept of Biocompatibility *Advances in Biomaterials Science and Biomedical Applications* ed R. Pignatello (Rijeka Croatia: In Tech) chapter 8 pp 207–40
[24] Chen Q Z, Li Y, Jin L Y, Quinn J M and Komesaroff P A 2010 *Acta Biomater.* **6** 4143–53
[25] Adams L A, Essien E R, Shaibu R O and Oki A 2013 *New J. Glass Ceram.* **3** 11–5
[26] Shahzadi P, Shahnaz A, Saadet M, Shah W A, Hussain T and Alvi M U 2014 *J. Chem. Mater. Res.* **2** 2–5
[27] Khalil E M A, El Batal H A and Hamdy Y M 2008 *Trans. Ind. Ceram. Soc.* 67 119–28
[28] Arepalli S K, Tripathi H, Vyas V K, Jain S, Suman S K, Pyare R and Singh S P 2015 *Mater. Sci. Eng.* C **49** 549–59
[29] Jurczyk K, Kubicka M M, Ratajczak M, Jurczyk M U, Niespodziana K, Nowak D M, Gajecka M and Jurczyk M 2016 *Trans. Nonferrous Met. Soc. China* **26** 118–25
[30] Goh Y F, Alshemary A Z, Akram M, Kadir A, Rafiq M, and Hussain R 2014 *Int. J. Appl. Glass Sci.* **5** 255–66
[31] Nezafati N, Moztarzadeh F, Hesaraki S, Mozafari M, Samadikuchaksaraei A, Hajibaki L and Gholipour M 2012 Effect of silver concentration on bioactivity and antibacterial properties of SiO_2-CaO-P_2O_5 sol-gel derived bioactive glass *Bioceramics* vol 23, ed E S Kayali, G Goller and I Akin (Istanbul: Trans Tech Publication) pp 74–9
[32] Vernè E, Di Nunzio S, Bosetti M, Appendino P, Brovarone C V, Maina G and Cannas M 2005 *Biomaterials* **26** 5111–9
[33] Peitl F O, Latorre G P and Hench L L 1996 *J. Biomed. Mater. Res.* **30** 509–14
[34] El Batal H A, Azooz M A, Khalil E M A, Soltan M A and Hamdy Y M 2003 *Mater. Chem.*

Phys. **80** 599–609
[35] Chajri S, Bouhazma S, Herradi S, Barkai H, Elabed S, Koraichi S I, El Bali B and Lachkar M 2015 *J. Mater. Environ. Sci.* **6** 1882–97
[36] Vallet-Regi M, Izquierdo-Barba I and Salinas A J 1999 *J. biomed. Mater. Res.* **46** 560–5
[37] Stanciu G A, Sandulescu I, Savu B, Stanciu S G, Paraskevopoulos K M, Chatzistavrou X, Kontonasaki E and Koidis P 2007 *J. Biomed. Pharm. Eng.* **1** 34–9
[38] Jurczyk M U, Jurczyk K, Miklaszewski A and Jurczyk M 2011 *Mater. Design.* **32** 4882–9
[39] Li S, De Wijn J R, Li J, Layrolle P and De Groot K 2003 *Tissue Eng.* **9** 535–48
[40] Griffith LG 2002 Ann. N.Y. Acad. Sci. **96** 83–95
[41] Gerhardt L C and Boccaccini A R 2010 *Materials* **3** 3867–910
[42] Pereira M M, Clark A E and Hench L L 1995 *J. Am. Ceram. Soc.* **78** 2463–8
[43] Saravanapavan P and Hench L L 2003 *J. Non-Cryst. Solids* **318** 14-26
[44] Li X, Wang X, He D and Shi J 2008 *J. Mater. Chem.* **18** 4103-9
[45] Bansal N P 1992 *J. Mater. Sci.* **27** 2922–33
[46] Vallet-Regi M, Arcos D and Pérez-Pariente J 2000 *J. Biomed. Mater. Res.* A, **51** 23–8
[47] Ruardy T G, Schakenraad J M, van der Mei H C and Busscher H J 1995 *J. Biomed. Mater. Res.* A **29** 1415–23
[48] Van Wachem P B, Hogt A H, Beugeling T, Feijen J, Bantjes A, Detmer J P and Van Aken W G 1987 *Biomaterials* **8** 323–8
[49] Roach P, Farrar D and Perry C C 2005 *J. Am. Chem. Soc.* **127** 8168–73
[50] Lampin M, Warocquier-Clérout R, Legris C, Degrange M and Sigot-Luizard M F 1997 *J. Biomed. Mater. Res.* A **36** 99–108
[51] Vogler E A 1998 *Adv. Colloid Interface Sci.* **74** 69–117
[52] De Rosa R L, Schader P A and Shelby J E 2003 *J. Non-Cryst. Solids* **331** 32–40

Effect of P, Na, Mg, and Ag content on the in vitro bioactivity, wettability and mechanical strength of sol-gel glasses

S Bouhazma[1], S Chajri[1], S Herradi[1], M Khaldi[1], A El Hachadi[2], B El Bali[1] and M Lachkar[1]

[1]Engineering Laboratory of Organometallic and Molecular Materials, Faculty of Sciences, University Sidi Mohamed Ben Abdellah, Po. Box 1796 (Atlas), 30000 Fez, Morocco.
[2]Laboratory of Process, Signals, Industrial Systems, Computing, High School of Technology, University Cadi Ayyad, Po. Box 549, Safi, Morocco.

Email: saraz2010@live.fr

Abstract. Bioactive glasses of the type SiO_2-CaO, SiO_2-CaO-P_2O_5, and SiO_2-CaO-P_2O_5-MO (M = Na, Mg, or Ag) were obtained by the sol-gel processing method. The obtained materials was characterized by X-ray powder diffraction (XRD), Fourier transform infrared spectroscopy (FTIR), and scanning electron microscopy coupled with energy dispersive spectroscopy (SEM/EDS). Contact angle and surface tension variation with time were determined at 25°C, respectively, by the sessile and pendant drop techniques, for distinct testing liquids: water, diiodomethane, formamide, and simulated body fluid (SBF). The in vitro studies showed that all gel-glasses compositions produced were bioactive. In the present work, three effects of elements (Mg, Na, Ag) in the glass were detected: (i) good mechanical strength with satisfactory biodegradability, (ii) formation of hydroxyapatite, which may promote good bone bonding and (iii) the good wettability.

1. Introduction

Many synthetic biomaterials have been studied as scaffolds to assist or enhance the body's own reparative capacity. Bioactive ceramics were attracting much attention for their excellent bone conductivity; these are reactive materials that can be integrated into the body. Glasses and glass-ceramics based on the SiO_2-CaO-P_2O_5 system constitute an important group of materials that have found wide applications in medicine as bone implants [1]. The sol-gel method is used to synthesize these bioglasses because it offers many advantages in comparison with the traditional fusion-quenching process. Furthermore, sol-gel glass exhibits high specific area and porous structures in the microporous or mesoporous range, which can promote cell adhesion, adsorption of biologic metabolites, and resorbability at controlled rates to match that of tissue repair [2-4]. Several studies have shown the relationship between hydrophobicity, surface tension and surface charge with the proteinaceous film formation and the cell/material interaction [5-7]. Cell attachment, adhesion and spreading influence the cells' capacity to proliferate and to differentiate, determining the failure or success of the implant in service [8].

The objective of the present investigation was to evaluate the in vitro bioactivity and the relation between the contact angle and mechanical strength of a novel sol-gel derived SiO_2-CaO-P_2O_5-MO (M = Na, Mg, and Ag). The properties of the elements-incorporated bioglass were compared with two bioglasses: SiO_2-CaO and SiO_2-CaO-P_2O_5. The last objective of this work was to demonstrate that incorporation of Na_2O, MgO, and Ag_2O has an effect on the bioactivity, wettability and mechanical properties.

2. Experimental procedure

The sol-gel synthesis of the glasses was performed as follows: initially, tetraethoxysilane (TEOS; Aldrich) was added to 0.1 M nitric acid and the mixture was allowed to react for 60 min for the acid hydrolysis of TEOS. Then a series of reagents were added in the following sequence, allowing 45 min for each reagent to react completely: triethylphosphate (TEP; Aldrich), calcium nitrate tetrahydrate (Aldrich), and other products: silver nitrate, sodium nitrate, magnesium nitrate and barium nitrate (Aldrich). After the final addition, mixing was continued for 1 h to allow the completion of hydrolysis. The resultant solution was kept in a sealed Teflon container for 3 days at ambient temperature to allow gelation to occur. The resultant gel was kept in a sealed container and heated at 70°C for an additional 3 days. The water was removed and a small hole was inserted in the lid to allow the leakage of gases while the gel was heated to 150°C for 48 hours to remove all the free water. The dried gel was then heated for 3h at 700°C to stabilize the glass and eliminate residual nitrates. The characterization of the bioglass samples was carried out using X-ray diffraction (XRD) analysis, Fourier transform infrared spectroscopy (FTIR), scanning electron microscope (SEM). The compositions of the glasses are listed in Table 1.

Table 1. Nominal composition (mol %) of as-prepared glasses.

Series	SiO_2	CaO	P_2O_5	Na_2O	MgO	Ag_2O
BG-b	77	23	0	0	0	0
BG-t	57	37	6	0	0	0
BG-qNa	57	18.5	6	18.5	0	0
BG-qMg	57	29.5	6	7.5	0	0
BG-qAg	57	32	6	5	0	0

2.1. Study of wettability

Wettability, by controlling the interaction with physiological fluids, would primarily influence cell behavior on biomaterials. Biomaterials first contact a proteinaceous liquid phase, almost aqueous in nature, leading to surface reorganization of proteins followed by cell attachment on biomaterials. Contact angle measurements were performed by using a goniometer by the sessile drop method (GBX-France). Two polar liquids (water and formamide) and one non-polar liquid (diiodomethane) were used. A drop of ultra-pure water, diiodomethane, formamide or SBF was placed on the glass disk mounted on a prepared substrate plate and the image was immediately sent by the camera to the computer for analysis. The contact angles of water were measured after 10 s. The temperature and the humidity were constant during the experiment (22°C and 63% respectively). The measurements were repeated by depositing at least three drops on each substrate, and the final results were presented by the average of the measured drops.

2.2. In vitro study
In order to evaluate bioactivity, we used the simulated body fluid (SBF) proposed by Kokubo *et al.* [9]. SBF is an acellular aqueous solution with an inorganic ion composition almost equal to human plasma. The bioactivity response of the materials was evaluated on the surface of powder samples (90 mg) after being soaked in 45 mL of SBF for different times.

2.3. FTIR analysis
The monitoring of powder modification after in vitro bioactivity tests was performed by FTIR using an IR Bruker VERTEX 70 spectrometer. The FTIR spectra were recorded from 400 to 4400 cm^{-1} with the resolution of 4 cm^{-1}.

2.4. Antibacterial activity
The antibacterial activity of samples was evaluated by the plate-counting method against E. coli (O128B12). Before the test, all samples were sterilized at 121°C for 20 min by autoclave. The concentration of bacterial solution was adjusted to 2.108 CFU/mL.

3. Results
The results are subdivided in two parts: the first one (Part A) is related to the synthesis and characterization of samples (in vitro study, characterization by RX, FTIR, and SEM). The second one (Part B) is related to the relation between the wettability and porosity for all glasses. Part A: The bioglasses phases were investigated by X-ray diffraction. Figure 1 shows the XRD pattern of bioglasses, which has been produced after thermal treatment on 700°C.

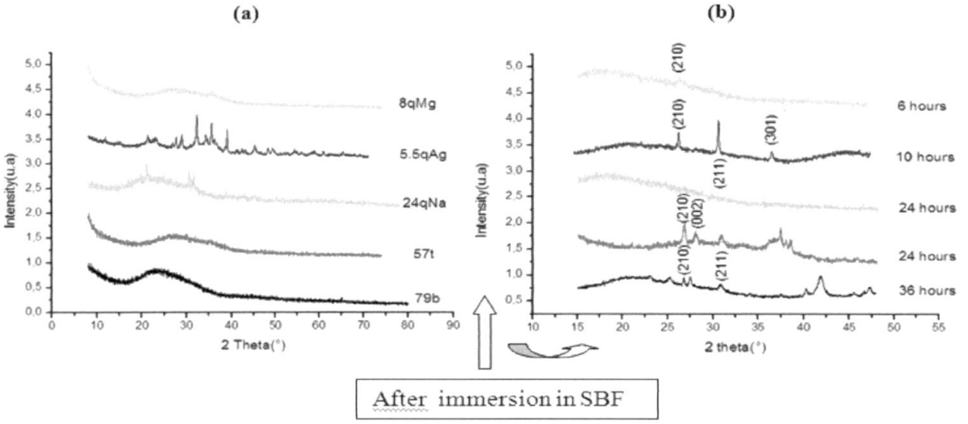

Figure 1. XRD spectra of sol-gel derived glasses after sintering at 700°C for 3 h.
(a): before soaking, (b): after soaking in SBF.

Some samples almost take amorphous state indicative of the internal disorder and glassy nature of these materials [10]. The first spectrum is the XRD spectra of the sol-gel derived and sintered 79b powders; it's amorphous for certain glasses. The Ca^{2+} released to SBF and the increase in pH are in agreement with the mechanism proposed by Kokubo for Hydroxide Carbonated Apatite (HCA) formation on the surfaces of bioactive glasses. In such glasses, an interchange between the Ca^{2+} ions of the glass and the H_3O^+ of the solution takes place. This gives rise to the formation of Si-OH groups on the glass surface, inducing apatite nucleation. Data obtained by XRD show that an apatite-like phase is formed on glass surface. Figure 1 shows XRD patterns obtained from the glass surfaces before and after soaking in SBF. XRD patterns of the glass before soaking did not show maxima corresponding to any crystalline phase, suggesting that our glass was amorphous or had crystallites too small to be detected by XRD. For 79b, 57t, and 24qNa, the patterns obtained for the glass after soaking for 1h, 6h, and 12h were similar to those obtained for the glass before soaking. X-ray diffraction patterns obtained for the glass after 1 days' soaking showed two wide diffraction maxima at 25 and 32°, corresponding to (210) and (211) reflections of an apatite-like phase. For 5.5qAg and 8qMg, after just 6h, the (211) and (210) and (310) reflections of the apatite phase appeared owing to the increase in its crystallinity. Data obtained by XRD show that an apatite-like phase is formed on glass surface. for binary, ternary and quaternary system doped with sodium, the broad-shape profiles of diffraction maxima for 24h and 36h point out that this phase is nanocrystalline, for other glasses (5.5qAg, and 8qMg), the bioactivity has improved, because 1 days' soaking showed two wide diffraction maxima corresponding to (210) and (211) reflections of an apatite-like phase.

Some glass systems such as SiO_2-CaO-P_2O_5, SiO_2-CaO-P_2O_5-Na_2O, SiO_2-CaO-P_2O_5-Ag_2O, and SiO_2-CaO-P_2O_5-MgO have also been investigated [11]. These results are confirmed by FTIR analysis. The major vibrational bands observed in the FTIR on bioglass sample indicated carbonated apatite formation on the disc surfaces. Figure 2 shows the FTIR spectra of sample before and after soaking in SBF for different periods. Before soaking in SBF, the FTIR spectra of the sample exhibited Si–O–Si stretching and bending bands. The band at around 1080 cm^{-1} corresponds to the vibrational mode of the asymmetric stretch of Si–O–Si, the band at 800 cm^{-1} corresponds to the symmetric stretch of Si–O, and the strong band at 475 cm^{-1} corresponds to the vibrational mode of the bending of Si–O–Si. After soaking at various times in SBF solution, additional peaks appeared in the FTIR spectra around 550–600 cm^{-1}. The twin bands at 570 cm^{-1} and 600 cm^{-1} (P–O bands) which correspond to antisymmetric vibration mode of P–O in amorphous calcium phosphate indicate apatite formation in SBF in addition the bands at 1490, 1426, and 872 cm^{-1} corresponding to carbonates [12].

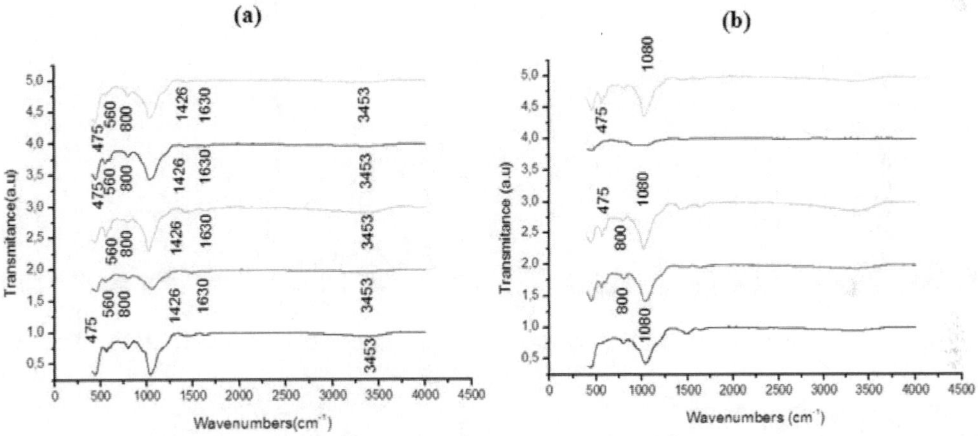

Figure 2. FTIR spectra of sol-gel derived glasses after sintering at 700°C for 3h.
(a): before soaking, (b): after soaking in SBF.

IOP Conf. Series: Journal of Physics: Conf. Series **984** (2018) 012012 doi:10.1088/1742-6596/984/1/012012

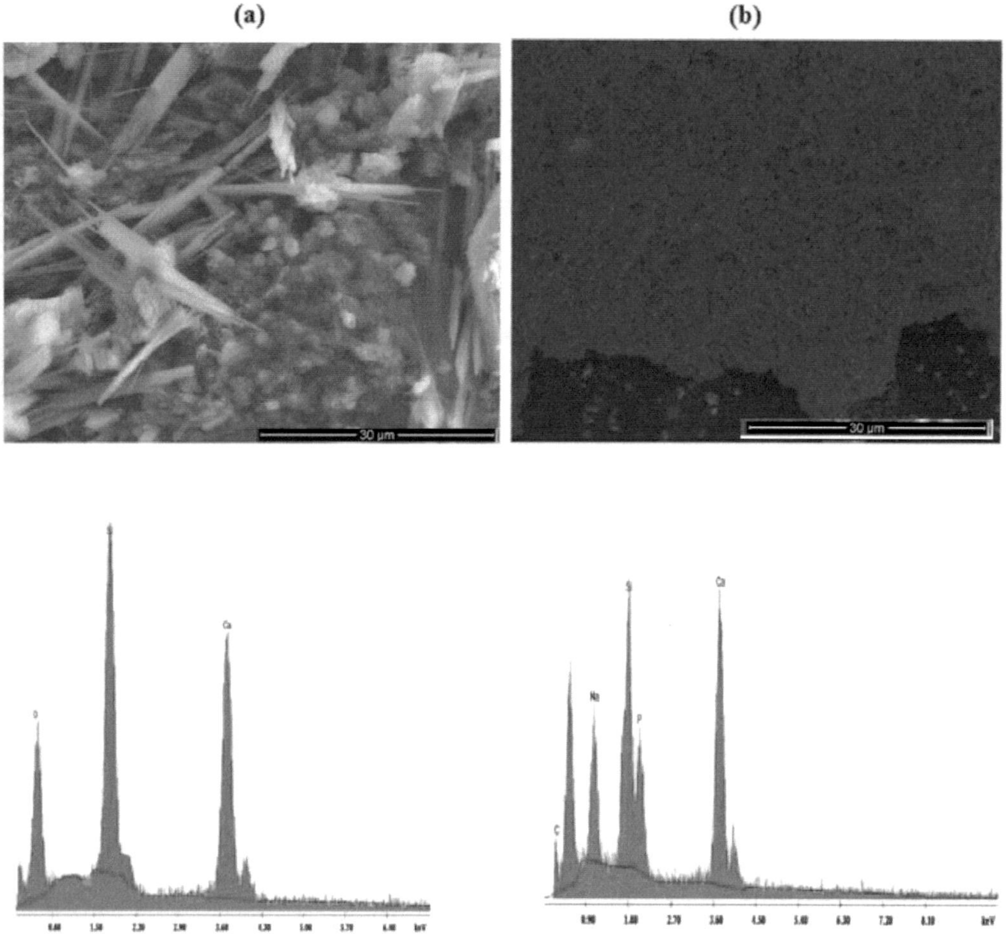

Figure 3. SEM micrographs and EDX patterns of the glass surface before and after 24h soaking in SBF for 24qNa glass.

Figure 3 shows the SEM micrographs and the EDX spectra of 24qNa sample surfaces after different soaking times in the SBF. The micrograph of the in situ sample shows a heterogeneous surface, typical in a glass, and the EDX study confirms that it is solely composed of oxygen, silicon, and calcium. After XRD and FTIR analysis the SEM observations provide evidence that two or more layers grew on the surface of the gel-glasses. The SEM micrographs of the 24qNa gel glass after 24 hours exposure reveal a difference in morphology within the formed HCA layer. A possible development of layers can be visualized in Figure 3b. The development of the formed HA layer on the surface of these materials can also be followed in the EDX spectra.

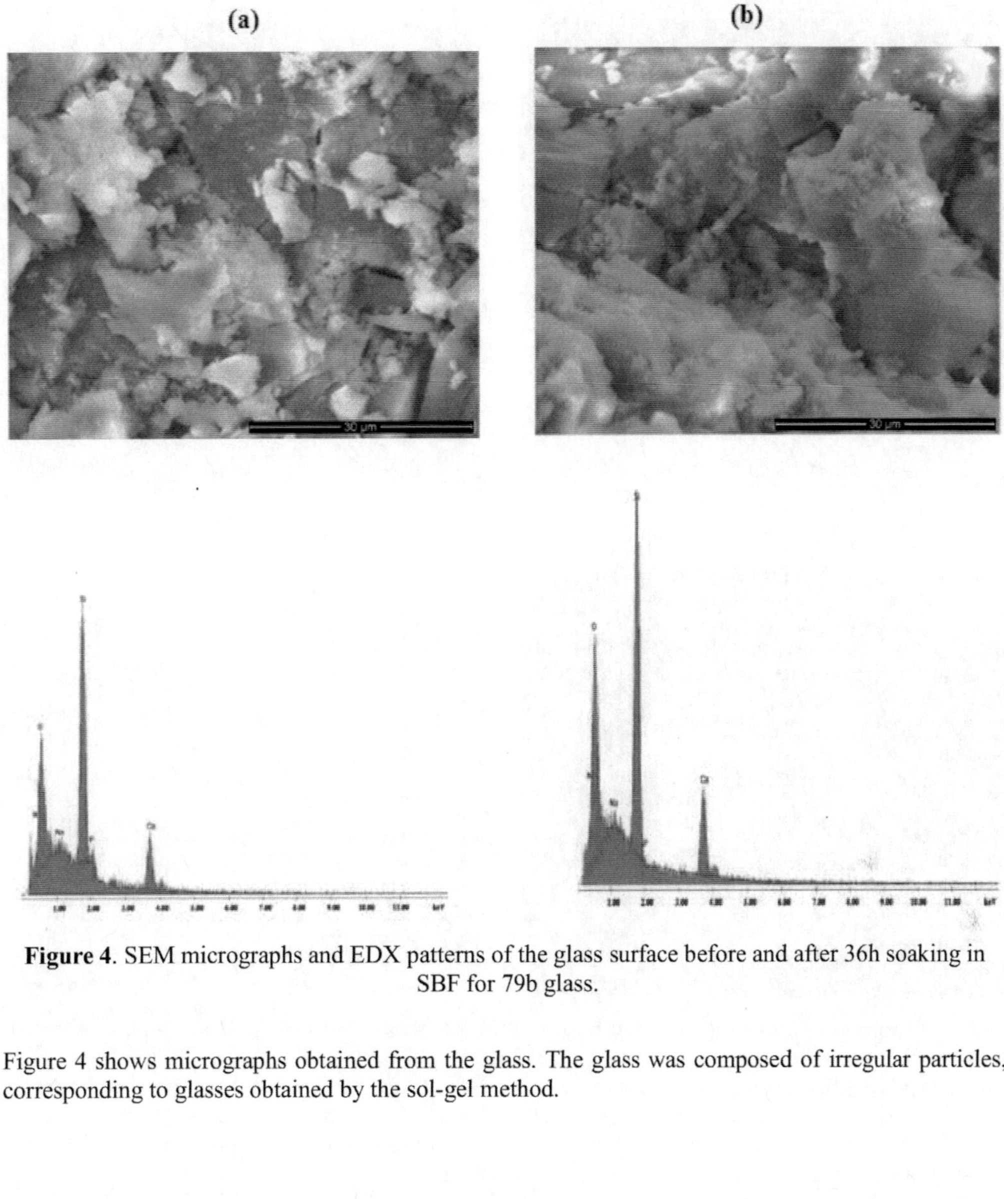

Figure 4. SEM micrographs and EDX patterns of the glass surface before and after 36h soaking in SBF for 79b glass.

Figure 4 shows micrographs obtained from the glass. The glass was composed of irregular particles, corresponding to glasses obtained by the sol-gel method.

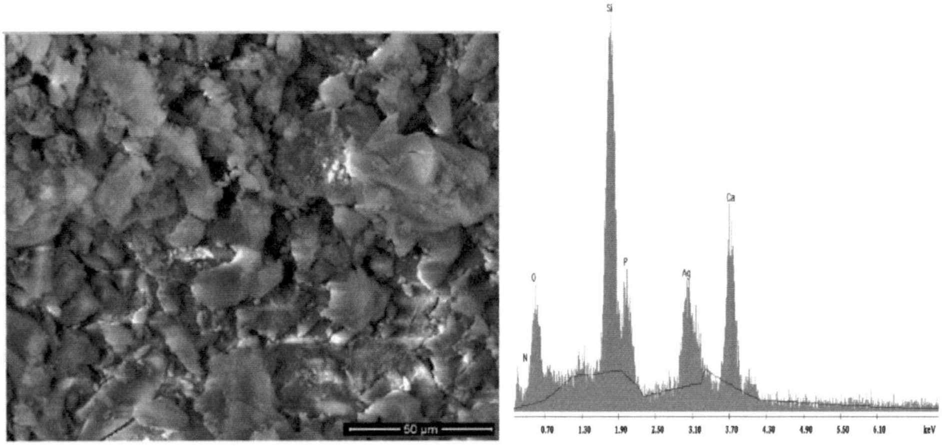

Figure 5. SEM micrographs and EDX patterns of the glass for 5.5qAg glass.

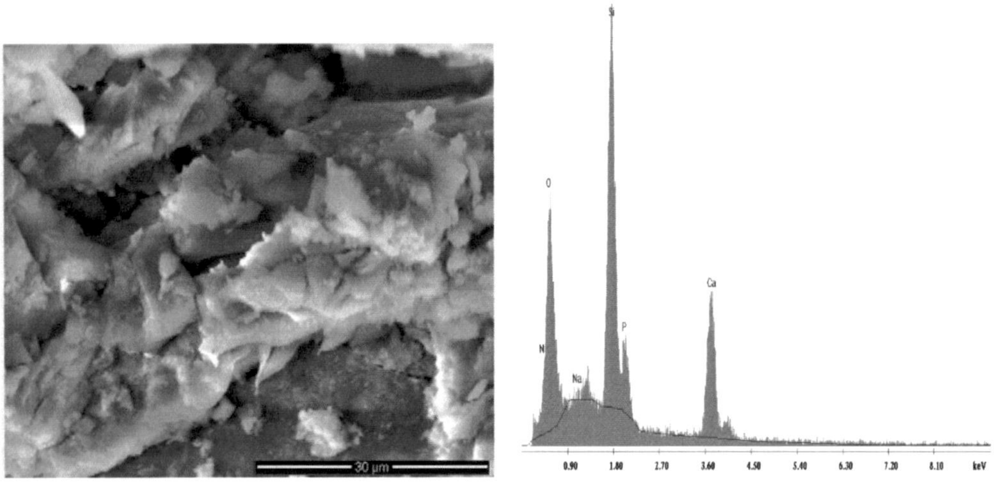

Figure 6. SEM micrographs and EDX patterns of the glass surface for 7.5qMg glass.

The element of silicon is also important because the HCA formation time is also dependent of the Si quantity. In order to yield apatite quickly, a bioactive glass must present a medium rate of SiO_2 (between 44 and 50%). When the quantity of Si is too low, the initial silicate network is more depolymerized, and the reaction requires a longer time in order to obtain a highly polymerized silica gel phase on which the amorphous calcium phosphate will precipitate. If the Si rate is too high, the highly polymerized network slows down the migration of phosphate groups, and delays the formation of HCA at the surface. Rehman *et al.* [13] showed that when the rate of silica increases in glasses, the network takes longer time to release its CaP groups, leading to a slower HCA formation. We must mention that the phosphorus-free glasses have already been studied in previous papers [14]. In this work, we have five different bioglasses with different compositions in the binary, ternary and quaternary systems. When phosphorus is added to the compositions, the two factors affecting the velocity of HCA formation are the Si and P percentages. It is known that phosphorus forms Na–Ca orthophosphate groups when added to the glass. The migration of these groups through the silica-rich layer improves the formation of the CaP phase at the surface of the material. For the bioglass doped with sodium it is known that the mechanical properties may be improved by incorporating Na_2O into bioactive glasses, which can result in the formation of a hard yet biodegradable crystalline phase from bioactive glasses when sintered. However, production of Na_2O-containing bioactive glasses by sol-gel methods has proved to be difficult. Our team already studied the influence of sodium addition into the glasses of the quaternary system [15]. A comparative study of sol-gel derived 79b and 57t (a Na_2O-free composition) and 24qNa glass ceramic materials revealed that the latter possesses a number of features desirable in biomaterials used for bone tissue engineering, including the crystalline phase $Na_2Ca_2Si_3O_9$ that couples good mechanical strength (part B) with satisfactory biodegradability, formation of hydroxyapatite, which may promote good bone bonding and cytocompatibility. For glass doped with magnesium, the element of Mg has shown that it play an important role in bone development and maintenance [16, 17]. Landi and Cai substituted magnesium for calcium in biological hydroxyapatite and indicated that magnesium-substituted apatite materials had excellent biocompatibility and great osteoconductivity over time [18, 19]. In addition, Vallet-Regi *et al.* [20] and Tahriri and co-workers [21] synthesized sol-gel derived glasses in the system CaO-MgO-P_2O_5-SiO_2 the objective in these studies is to study if the presence of MgO, in this family of bioactive glasses, affects the rate of formation of the hydroxycarbonate apatite layer, the thickness of such a layer and the possible formation of secondary phases which compete in stability with the apatite-like phase.

In this work, we report the synthesis of sol-gel glass in the system CaO-MgO-P_2O_5-SiO_2 with 7.5 mol% MgO content at 700°C. The pattern of sample obtained at 700°C corresponds to amorphous materials; no crystalline phases were detected, but X-ray diffraction patterns obtained for the glass after 6 hours soaking showed two diffraction at 26° and 32°, corresponding to (210) and (211) reflections of an apatite-like phase. The mechanism of the antibacterial activity of the materials that contain Ag ions has been studied by several scholars [22, 23]. With 5% of Ag, the bioglass BG-qAg possesses a significant bactericidal activity. We can suggest that bioglass BG-qAg markedly inhibits the growth of the gram-negative bacteria when tested at concentrations in the range of 0.06 to 1 mg/mL. It exhibits a significant bacteriostatic effect for cultures containing excess of 0.3 mg/mL of the sample. In conclusion, this work present the effect of each element for the bioglass studied.

Part B: Surface energetics, by controlling the interaction with physiological fluids, would primarily influence cell behavior on biomaterials. The best wetting properties to give a picture of biomaterials behavior are the contact angle and the work of adhesion [24]. The contact angle of water on the all glasses is significantly lower than the lowest value (48°) published for a set of other bioceramics, namely HA, β-TCP and glass reinforced HA composites in bulk and film forms [25, 26]. In this part we show the relation between the contact angle and mechanical strength. The overall set of results from contact angle measurements on BG-b, BG-t, BG-qNa, BG-qMg, and BG-qAg materials are presented successively in Table 2, Table 3, Table 4, Table 5, and Table 6. The contact angle of the SBF liquid was measured allowing a comparison to that of the water.

Table 2. Contact angle for several liquids on BG-b.

Liquid	**Contact angle (°)**
Water	15.16 ± 2
Diiodomethane	18.65 ± 2
Formamide	19.03 ± 2
Simulated body fluid (SBF)	17.76 ± 2

The contact angles of the different surfaces studied are under the theoretical limit of 65°. In this study, contact angle measurements showed distinct differences between the glass substrates, revealing a stronger dependence on the surface chemical composition. The BG-b bioglass is hydrophilic, with a contact angle of 15.16 ± 2°, BG-b is significantly much more hydrophilic. The hydrophobicity is the hard relation on the porosity and mechanical strength.

Table 3. Contact angle for several liquids on BG-t.

Liquid	Contact angle (°)
Water	21.2 ± 2
Diiodomethane	18.0 ± 2
Formamide	10.8 ± 2
Simulated body fluid (SBF)	17.4 ± 2

Despite doping with phosphorus, the values of contact angle did not undergo a great change; bioglass BG-t is hydrophilic comparing with the BG-b which remains more hydrophilic.

Table 4. Contact angle for several liquids on BG-qNa.

Liquid	Contact angle (°)
Water	11.7 ± 2
Diiodomethane	13.9 ± 2
Formamide	11.9 ± 2
Simulated body fluid (SBF)	10.7 ± 2

After doping with sodium, it is found that the contact angle is decreased (11.7 ± 2) and the glass became more hydrophilic with adequate strength.

Table 5. Contact angle for several liquids on BG-qMg.

Liquid	Contact angle (°)
Water	10.16 ± 2
Diiodomethane	13.65 ± 2
Formamide	12.03 ± 2
Simulated body fluid (SBF)	15.76 ± 2

The effect of MgO on the glass ceramics is the development of the formation of apatite [1,27]. The role of MgO in the surface behavior of bioglasses is contradictory, because it appears in the glass structure as a modifier or as a network former or as an intermediate. Hence, appropriate in vitro behavior of MgO in silicate glasses is required even though it is found to inhibit mineralization in bones [28]. Recent investigations on the SiO_2-CaO-P_2O_5-MgO systems how that high MgO content can promote in vitro apatite formation [29]. However, the information on the mechanism of apatite formation with different compositions of MgO and its influence on in vitro behavior are scanty [30].

Table 6. Contact angle for several liquids on BG-qAg.

Liquid	**Contact angle (°)**
Water	39.64 ± 2
Diiodomethane	23.15 ± 2
Formamide	20.8 ± 2
Simulated body fluid (SBF)	30.69 ± 2

The historic use of silver foils in surgical treatment of wounds and broken bones is documented [31]. The general finding is that bacteria show a low propensity to develop resistance to silver-based products, and therefore both metallic and ionic silver have been incorporated into several biomaterials such as polyurethane, hydroxyapatite (HAP) and bioactive glasses [32]. Silver-containing products are also interesting materials for wound repair applications [33]. In addition to its antimicrobial activity, the silver bioglass has a high mechanical strength and an exemplary contact angle, and in the literature there are no studies who are interested in the study of the contact angle of such an example.

Table 7. Mechanical property of glasses.

Samples	BG-b	BG-t	BG-qMg	BG-qNa	BG-qAg
Compressive strength values (MPa)	86.25	95.13	80.34	60.18	110.23

Although the effect of the contact angle on the mechanical properties of bioglasses is important, there are no studies in the literature which is interested in the relationship between the contact angle and strength. In this work we can see that the higher the bioglass is more hydrophilic it is less resistant. For example, bioglass BG-qAG is the strongest because it has a value of the contact angle more grade of others (Table 7). Comparing with the literature, it is clear that our glasses are hydrophilic which promotes cell adhesion and allowing the exchange of ions in human plasma.

4. Conclusion

Fine powders of binary, ternary, and quaternary system have been successfully synthesized using the sol-gel technique in aqueous solution under ambient conditions. The samples prepared exhibited the high bioactivity, high wettability, and high mechanical strength. The BG-qAg bioglass can be good candidate because the composition presents a medium quantity of SiO_2, a high rate of P_2O_5 and a reasonable quantity of Ag_2O. All bioglasses in this work present a good hydrophilic behavior permitting cell adhesion and a good mechanical strength.

Acknowledgements- The authors would like to acknowledge the support and technical assistance of Interface Regional University Center (University Sidi Mohammed Ben Abdellah, Fez, Morocco).

References

[1] Balamurugan A, Balossier G, Michel J, Kannan S, Benhayoune H, Rebelo A H S and Ferreira J M F 2007 *J. Biomed. Mater. Res. B: Appl. Biomater.* 83B 546
[2] Sepulveda P, Jones J R, and Hench L L 2002 *J Biomed. Mater. Res.* 59 340
[3] Hamadouche M, Meunier A, Greenspan D C, Blanchat C, Zhong J P, La torre G P and Sedel L 2001 *J Key Eng. Maters* 593192
[4] Lenza R F S, Vasconcelos W L, Jones J R and Hench L L 2002 *J. Mater. Sci.Mater.* Med. 13
[5] Lampin M., Warocquier-Cl!erout R, Legris C, Degrange M and Sigot- Luizard M F 1997 *J Biomed Mater Res* 36 99
[6] Dewez J L, Schneider Y J, Rouxhet P G 1996 *J Biomed Mater Res* 30 373
[7] Hallab N J, K Bundy, O'Connor R, Clark R and Moses R L 1995 *J. Long-Term Effects Med. Implants* 3 209
[8] Anselme K 2000 *J. Biomaterials* 7 667
[9] Kokubo T, Kushitani H, Sakka S and Yamamuro T J 1990 A-W. *J. Biomed. Mater. Res.* 24 721
[10] Perez-Pariente J, Balas F, Roman J, Salinas A J and Vallet-Regi M 1999 *J. Biomed. Mater. Res.* 47 170
[11] Qi-Zhi Chen, Yuan Li, Li-Yu Jin, Julian Quinn M W and Komesaroff P A 2010 *J. Acta Biomater.* 6 4143
[12] Vidueau J and Dupius V 1991 *Eur. J. Solid State Inorg. Chem.* 28 303
[13] Rehman I, Hench L L, Bonfield W and Smith R 1994 J. Biomaterials 15 865
[14] Duée C, Désanglois F, Lebecq I, Moreau G, Leriche A and Follet-Houttemane C 2009 J. Non-Cryst. Solids 355 943
[15] Bouhazma S, Chajri S, Barkai H, Elabed S, Ibnsouda Koraichi S, El Bali B and Lachkar M 2015 Mor. J. Chem. 3 19
[16] Oliveira J M, Correia R N, Fernandez M H and Rocha J 2000 J. Non-Cryst. Solids 221
[17] Wu C T, Chang J, Wang J Y, Ni S Y and Zhai W Y 2005 J. Biomaterials 26 2925
[18] Landi E, Logroscino G, Proietti L, Tampieri A, Sandri M and Sprio S 2008 J. Mater. Sci. Mater. Med. 19 239
[19] Cai Y L, Zhang S, Zeng X T, Wang Y S, Qian M and Weng W J 2009 J. Thin Solid Films 517 5351
[20] Vallet-Regi M, Salinas A J, Roman J and Gil M 1999 *J. Mater. Chem.* 9 515
[21] Saboori A, Sheikhi M, Moztarzadeh F, Rabiee M, Hesaraki S, Tahriri M and Nezafati N 2009 *J. Adv. Appl. Ceram.* 108 155
[22] Bir F, Khireddine H, Touati A, Sidane D, Yala S and Oudadesse H 2012 *J. Appl. Surf. Sci.* 258 7021
[23] Eckhardt S, Brunetto P.S, Gagnon J, Priebe M, Giese B and Fromm K M 2013 *J. Chem. Rev.* 113 4708
[24] Vogler E A. 1993 Interfacial chemistry in biomaterials science (Wettability, surfactant science series, vol. 49) In Berg JC editor (Washington: University of Washington) Chapter 4 pp. 183-250

[25] Lopes M A, Monteiro F J, Santos J D, Serro A P and Saramago B 1999 *J. Biomed. Mater. Res.* 45 370
[26] Ferraz M P, Monteiro F J, Serro A P and Saramago B, 2001 *J. Biomaterials* 22 3105
[27] Ma J, Chen C Z, Wang D G, Shao X, Wang C Z and Zhang H M 2012 *J. Ceram. Int.* 38, 6677
[28] Kasuga T, Nakagawa K, Yoshida M and Miyade E 1987 *J. Mater. Sci.* 22 3721
[29] Oliveira J M, Correia R N and Fernandes M H 2002 *J. Biomaterials* 23 371
[30] Ni S, Chang J and Chou L 2008 *J. Mater. Sci. Mater. Med.* 19 359
[31] George N, Faoagali J, Muller M and Silvazine M 1997 *J. Burns* 23 493
[32] Bosetti M, Masse A, Tobin E and Cannas M 2002 *J. Biomaterials* 23 887
[33] Deitch E A, Marino A A, Malakanok V, Albright J A 1987 *J. Trauma.* 27 301

Synthesis, crystal structure and thermal study of the hybrid nickel sulfate: $C_6N_2H_{16}[Ni(H_2O)_6(SO_4)_2].2H_2O$

F A I Ngopoh[1], N Hamdi[1], S Chaouch[1], M Lachkar[1], I da Silva[2], B El Bali[1]

[1]Engineering Laboratory of Organometallic and Molecular Material, Faculty of Sciences, University Sidi Mohamed Ben Abdellah, P.O. Box 1796 (Atlas), 30000 Fez, Morocco
[2]ISIS Facility, STFC Rutherford Appleton Laboratory, Chilton, Oxfordshire OX11 0QX, UK

Email: atipof@yahoo.fr

Abstract. A new inorganic-organic hybrid open framework nickel sulfate $C_6N_2H_{16}[Ni(H_2O)_6(SO_4)_2].2H_2O$ has been synthesized by slow evaporation in aqueous solution using trans-1,4-diaminocyclohexane as structure-directing agent. It was characterized by single-crystal X-ray diffraction, infrared spectroscopy and analyzed by TGA-DSC. The compound crystallizes in the monoclinic space group $P2_1/n$, with the unit cell parameters of a = 6.2586 Å, b = 12.3009 Å, c = 13.2451 Å, β= 98,047°, Z= 4. Its crystal structure consists of isolated polyhedrons $[Ni(H_2O)_6]^{2+}$ and $[SO_4]^{2-}$ and free water which connects through hydrogen bonds. This association results in the porous framework where the protonated organic molecule trans-1,4-diaminocyclohexane is located as a counter ion. The IR spectra Shows the bands corresponding to the sulfate anion, water molecule and diprotonated trans-1-4-diaminocyclohexane. Thermal study indicates the loss of water molecules and the degradation of trans-1-4-diaminocyclohexane.

Keywords: Hybrid Ni-sulfate, Crystal structure, X-ray diffraction, Infrared Spectroscopy, Thermal analysis.

1. Introduction

During the last two decades, considerable efforts have been directed towards the synthesis of the hybrid organic-inorganic due to physical properties such as porosity, magnetism and electronic behavior that may lead to many applications in fields like: catalysis, gas absorption, non-linear optical and storage energy [1-4]. Among these, metal sulfates form an important class with known applications in optic and catalysis [5-12]. A large number of these materials are synthesized in the presence of organic amines acting as structure-directing agent and their crystal structure is governed by weak energy bonding, exhibiting the supramolecular open framework [13-15].

In the present work we report on the synthesis, crystal structure, infrared spectroscopy and the thermal properties of a new inorganic-organic nickel sulfate: $C_6N_2H_{16}[Ni(H_2O)_6(SO_4)_2].2H_2O$, with trans-1,4-diaminocyclohexane acting as structure-directing agent.

2. Experimental

2.1. Synthesis

The title compound was synthesized at 293 K by dissolving 400 mg of nickel sulfate $NiSO_4.6H_2O$ (Prolabo) in an aqueous solution (10 mL H_2O, 300 mg of H_3PO_2) of hypophosphorous acid (30-32%; sd Fine-Chem Limited). To the resulting solution, 70 mg of trans-1,4-diaminocyclohexane (Sigma-Aldrich) was added under continuous stirring. The mixture was allowed to concentrate at room temperature for one week and the investigated sample material was filtered off and washed with acetone. $C_6N_2H_{16}[Ni(H_2O)_6(SO_4)_2].2H_2O$ crystallizes as green crystals of plate shape.

2.2. Infrared spectrum

Infrared spectrum of the title compound was recorded on a VERTEX 70 FTIR Spectrometer, in the range 4000-400 cm^{-1}, using the ATR technique.

2.3. Crystal structure determination

Single-crystal X-ray diffraction measurement was carried out at room temperature using an Agilent Gemini S diffractometer equipped with a CCD detector and Mo radiation source. Acquired data was processed with the CrysAlisPro software [16]. Using Olex2 [17], the structure was solved with the olex2.solve [18] structure solution program using Charge Flipping and refined with the olex2.refine [18] refinement package using Gauss-Newton minimisation. All non-hydrogen atoms were refined anisotropically and hydrogen atoms were included in the model at calculated positions, refined with a rigid model with their Uiso value to 1.2Ueq of their parent atoms.

Table 1. Summary of crystal data and structure refinement.

CCDC no.	1588709
Chemical formula	$C_6N_2H_{16}[Ni(H_2O)_6(SO_4)_2].2H_2O$
Formula weight	511.15
Crystal system	Monoclinic
Space group	$P\,2_1/n$
a, Å	6.2586(3)
b, Å	12.3009(5)
c, Å	13.2451(6)
β, deg	98.047(5)
Volume, $Å^3$	1009.65(8)
Z	4
ρ_{calc} g cm^{-3}	1.6812
F(000)	541.6
μ(Mo Kα), mm^{-1}	1.244
Reflns (collected / unique, (Rint))	4028 / 2185 / 0.0274
Data / restraints / params	2185 / 2 / 157
R_1, wR [$F^2 > 2\sigma(F^2)$]	0.0343 / 0.0862
Goodness-of-fit on F^2	1.050
largest diff peak and hole (e $Å^{-3}$)	0.71 / -0.49

2.4. Thermal analysis

Thermogravimetric analysis (TGA) data were recorded on an SDT-Q600 analyzer from TA instruments. The temperature varied from RT to 900°C at a heating rate of 10 °C.min^{-1}. Measurements were carried out on samples in open platinum crucibles under air flow.

3. Results and discussions

3.1. Spectroscopy

The infrared spectrum of $C_6N_2H_{16}[Ni(H_2O)_6(SO_4)_2].2H_2O$ (Figure 1) shows bands corresponding to the vibrations of trans-1,4-diaminocyclohexane molecule, water molecule and sulfate anion. The broad band that appears at 3217.02 cm^{-1} is due to stretching vibrations of NH_2 group and the band at 2923.83 cm^{-1} is assigned to C-H stretching vibration [19]. All these bands indicate the presence of trans-1,4-diamonocyclohexane in the structure. The bands observed at 3102.32 and 1649.41 cm^{-1} are associated to the stretching and bending vibrations of OH group [19]. This confirms the presence of crystal water. The large band at 1043.82 cm^{-1} is attributed to the S-O stretching vibration of SO_4 group [20].

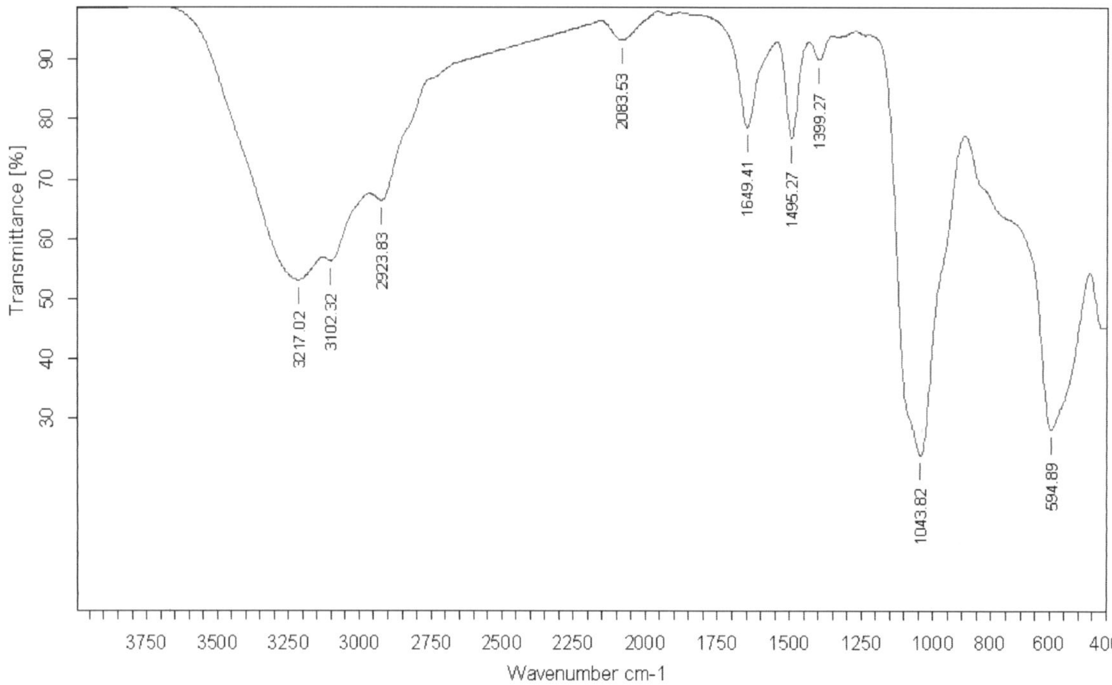

Figure 1. IR spectrum of $C_6N_2H_{16}[Ni(H_2O)_6(SO_4)_2].2H_2O$

3.2. Description of the crystal structure

Single-crystal structure analysis reveals that the asymmetric unit of title compound contains fourteen crystallographically independent non-hydrogen atoms, including one nickel atom, one sulfur atom, one nitrogen atom, three carbon atoms and eight oxygen atoms. Figure 2 depicts a projection of the crystal structure of the title compound on the crystallographic plane (001). Ni atom lies on an inversion center, it is coordinated to six water molecules to form $[Ni(H_2O)_6]^{2+}$ octahedron. The equatorial bond lengths Ni-O9 and Ni-O10 are 2.084 and 2.026 Å respectively while the axial bond length is 2.028 Å. The Average <Ni-O> bond length is 2.038 Å, these distances are very close to those found in [S-$C_5H_{14}N_2$][$Ni(H_2O)_6$]$(SO_4)_2$ (2.0518-2.0732 Å) [21]. The deformation of $[Ni(H_2O)_6]^{2+}$ octahedron can be evaluated by calculating the bond length distortions according to the equation (1) [22-23].

$$\Delta_{oct} = \frac{1}{6}\sum([Ri - \bar{R}]/\bar{R})^2 \quad (1)$$

Where R_i is the individual Ni-O bond length and $\bar{R}$ is the average Ni-O bond length.

A value of 1.72610^{-3} indicates thus a moderate distortion of the octahedron $[Ni(H_2O)_6]^{2+}$. The S atom, located in general position, is coordinated to four oxygen atoms forming a free $[SO_4]^{2-}$ tetrahedron, which is not coordinated to metal atom. The S-O distances vary from 1.463 to 1.483 Å, behaving the same as in $[S\text{-}C_5H_{14}N_2][Ni(H_2O)_6](SO_4)_2$ (1.462-1.481 Å) [20]. The polyhedrons $[Ni(H_2O)_6]^{2+}$ and $[SO_4]^{2-}$ and free water molecules are linked to each other by O-H---O hydrogen bonds, to form a two dimensional anionic inorganic supramolecular open network . The charge of inorganic network is compensated by the protonated trans-1,4-diaminocyclohexane molecules $[C_6N_2H_{16}]^{2+}$ which are held in the pores through N-H---Os interactions (Os from to sulfate anions) and giving rise to an organic-inorganic supramolecular open framework (Figure 3). This framework is comparable to crystal structure of $(C_7N_2H_7)_2[Ni(H_2O)_6(SO_4)_2].4H_2O$ [24].

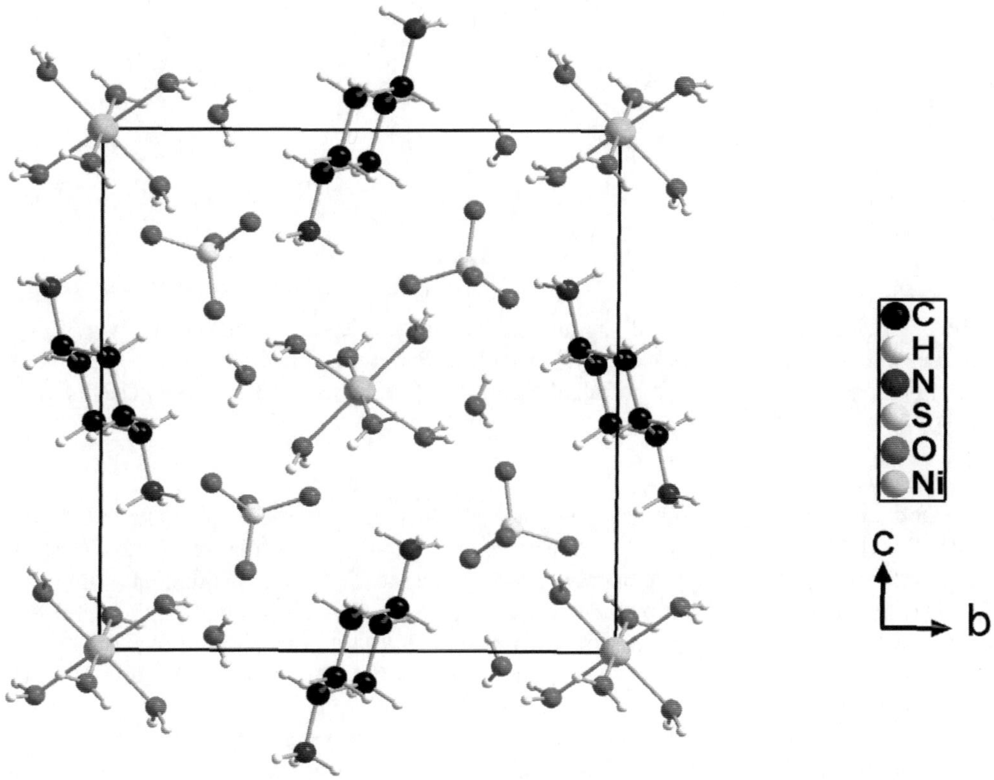

Figure 2. projection of the crystal structure of $C_6N_2H_{16}[Ni(H_2O)_6(SO_4)_2].2H_2O$ on the crystallographic plane (001).

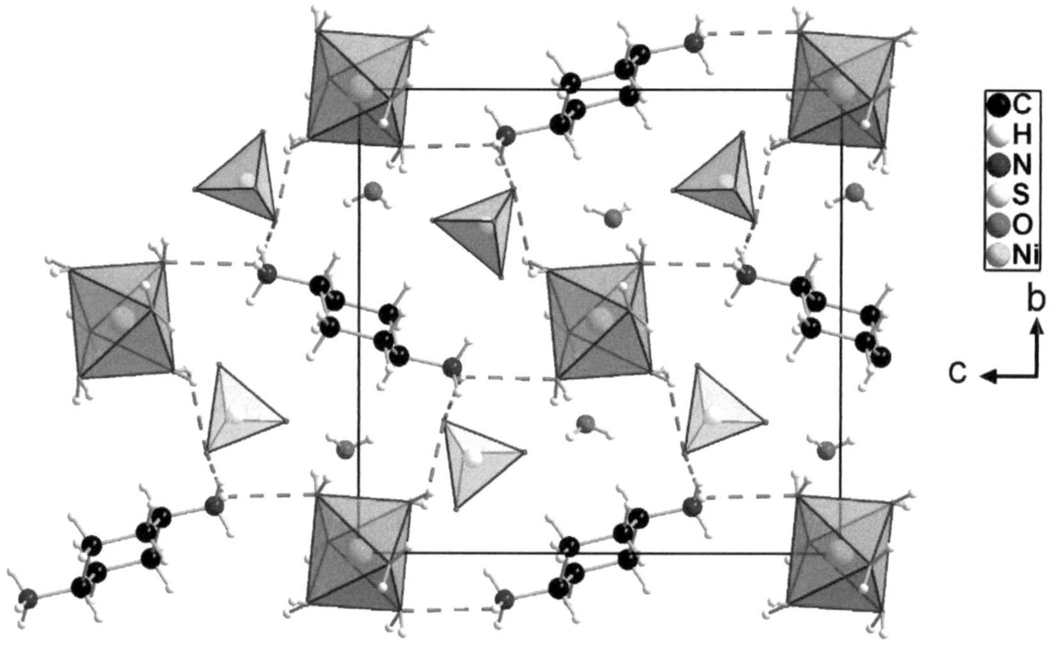

Figure 3. A fragment of the hybrid supramolecular in the title compound view along [100] direction, with the trans-1,4-diaminocycloxehane located in the pores. The green and yellow polyhedral represent respectively $[Ni(H_2O)_6]^{2+}$ octahedra and $[SO_4]^{2-}$ tetrahedra. H-Bonds as Dashed lines.

3.3. Thermal analysis

The results of the TGA-DSC analyses are plotted on the Figure 4. The TGA analysis of $C_6N_2H_{16}[Ni(H_2O)_6(SO_4)_2].2H_2O$ shows five steps of weight loss in a total of 70.55%. The first weight loss of 24.55%, starting at room temperature and ending at 150°C is in agreement with the removal of 6.5 water molecules per unit formula (theoretical loss, 22.9%). Two endothermic thermo-differential signals are associated with this dehydration reaction. It means that the removal of these 6.5 water molecules is done in two steps but very difficult to distinguish like shown on the TGA curve. The second weight loss of 5.039% in range of 150-200°C is attributed to the loss of 1.5 water molecules (theoretical loss, 5.329%). This second dehydration reaction is associated to an exothermic thermo-differential signal. The third weight loss of 23.69% occurred between 250-400°c correspond to the decomposition of trans-1,4-diaminocyclohexane molecule (theoretical loss, 22.705%). As one can expect the disintegration of organic content follows a discontinuous three-step process. According DSC curve this pyrolytic reaction is characterized by three exothermic signals. The weight loss of 7.062% and 9.939% in range of 400-800°C attributed to the final degradation of 2D inorganic framework, it can be due to the release of one dioxide molecule (theoretical loss, 6,7%) and one dioxide sulfur molecule (theoretical loss, 12%). This thermal decomposition is associated with two exothermic thermo-differential signals.

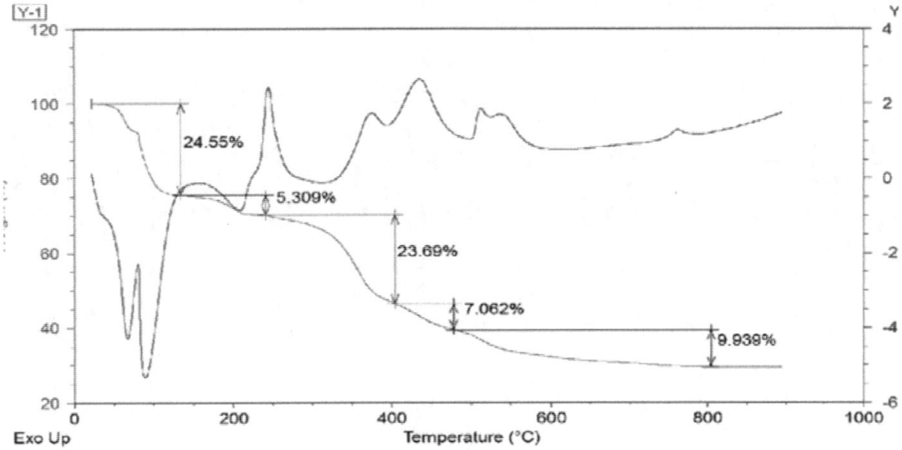

Figure 4. Thermal analysis of title compound. With TG (green), DSC (blue) signals

4. Conclusion

The 2D-hybrid nickel sulfate with trans-1,4-diamonocyclohexane molecule acting as structure-directing agent has been synthetized by slow evaporation. Its crystal structure exhibits a supramolecular layer made up by isolated entities $[Ni(H_2O)_6]^{2+}$, $[SO_4]^{2-}$, H_2O and $[C_6N_2H_{16}]^{2+}$ which are held together through O-H---O and N-H---O hydrogen bonds. The DSC-TG analysis shows that beyond 50°C this compound decomposes. The infrared bands that appear at 3217.02, 3102.32, 2923.83 and 1043.82 cm^{-1} are associated to the stretching vibrations of O-H, N-H, C-H and S-O bonds respectively and confirms the presence of $[SO_4]^{2-}$, H_2O and $[C_6N_2H_{16}]^{2+}$ entities in crystal structure of the tittle compound.

5. Acknowledgments

BEB would thank Prof. Aurelio Cabeza Díaz, Departamento de Química Inorgánica, Cristalografía y Mineralogía, Facultad de Ciencias, Universidad de Málaga (SPAIN) for the thermal measurements

References

[1] Sanchez C, Belleville P, Popall M and Nicole L 2011 *Chem. Soc. Rev.* **40** 696
[2] Dolbecq A, Dumas E, Mayer C and Mialane P 2010 *Chem. Rev.* **110** 6009
[3] Sanchez C, Ribot F and Lebeau B 1999 *J. Mater. Chem.* **9** 35
[4] Judeinstein P and Sanchez C 1996 *J. Mater. Chem.* **6** 511
[5] Morimoto C, and Lingafelter E 1970 *Acta cryst.* B **26** 335
[6] Al-Dajani M, Wahab H, Mohamed N, Goh J and Fun H 2010 *Acta Cryst.* E **66** m917
[7] Fleck M, Bohaty L and Tillmanns E 2004 Solid State **6** 469
[8] Rekik W, Naili H, Mhiri T and Bataille T 2011 *Acta Cryst.* E **67** m1176
[9] Doran M, Norquist M and O'Hare D 2005 *Acta Cryst.* E **6** m881
[10] Yahyaoui S, Rekik W, Naili H, Mhiri T and Bataille T 2007 *J. Solid State. Chem.* **180** 3560
[11] Tiwari R, Kumar J and Behera J 2016 *Chem. Commun.* **52** 1282
[12] Czpla Z, Komar J, Marciniszyn T, Poprawski R 2011 Solid State Ionics **196** 25
[13] Rekik W, Naili H, Bataille T, Roisnel T and Mhiri T 2006 *Inorg. Chim. Acta* **359** 3954
[14] Rekik W, Naili H, Mhiri T and Bataille T 2008 Materials Research Bulletin **43** 2709
[15] Naili H, Rekik W, Bataille T and Mhiri T 2006 Polyhedron **25** 3543
[16] CrysAlisPro Software System, Version 1.171.38.41 ;Rigaku Corporation : Oxford, U.K., 2016
[17] Dolomanov O, Bourhis L, Gildea R, Howard J and Puschmann H 2009 *J. Appl. Cryst.* **42** 339
[18] Bourhis L, Dolomanov O, Gildea R, Howard J and Puschmann H 2015 *Acta. Cryst.* **A71** 59

IOP Conf. Series: Journal of Physics: Conf. Series **984** (2018) 012013 doi:10.1088/1742-6596/984/1/012013

[19] Stuart, B 2005 Infrared Spectroscopy:Fundamentals and Applications (John Wiley & Sons)
[20] Nakamoto K 1986 Infrared and Raman Spectra of Inorganic and Coordiantion compounds (New York : Wiley)
[21] Hajlaoui F, Naili H, Yahyaoui S, Norquist A, Mhiri T and Bataille T 2012 *J. Organomet. Chem.* **700** 110
[22] Robinson K, Gibbs G and Ribbe P 1971 Science **172** 567
[23] Fleet M 1976 *Min. Mag.* **40** 531
[24] Kammoun O, Rekik W, Naili H and Bataille T 2015 *New J. Chem.* **39** 2682

$(C_6N_2H_{16})[Co(H_2O)_6](SO_4)_2.2H_2O$: A new hybrid material based on sulfate templated by diprotonated trans-1,4-diaminocyclohexane

N Hamdi[1], F A I Ngopoh[1], I da Silva[2], B El Bali[1] and M Lachkar[1]

[1] Engineering Laboratory of Organometallic and Molecular Materials (CNRST, URAC 19), Faculty of Sciences, University Sidi Mohamed Ben Abdellah ,Po. Box 1796 (Atlas), 30000 Fez, Morocco
[2] ISIS Facility, STFC Rutherford Appleton Laboratory, Chilton, Oxfordshire OX11 0QX, UK

Email: hamdi.najlae@gmail.com

Abstract. Employing trans-1,4-diaminocyclohexane (DACH) as template, the new hybrid sulphate $(C_6N_2H_{16})[Co(H_2O)_6](SO_4)_2.2H_2O$ was prepared in solution. Single-crystal X-ray diffraction analysis shows that it crystallizes in the monoclinic system (S.G.: P 21/n), with the following unit-cell parameters (Å,°): a = 6.2897(2), b = 12.3716(6), c = 13.1996(4), β = 98.091(3) V = 1016.89(7) Å3, Z = 4. Its 3D crystal structure is made upon isolated [Co(H2O)6] octahedra, regular [SO4] tetrahedra, protonated DACH and free H2O molecules, which interact through N-H···O and O-H···O hydrogen bonds. The Fourier transform infrared result exhibits bands corresponding to the vibrations of DACH, sulfate group and water molecules. The thermal decomposition of the phase consists mainly in the loss of the organic moiety and one sulfate group, leading thus to the formation of anhydrous cobalt sulfate.

Keywords: Sulfate; Crystal Structure; Hybrid; FTIR; Thermal behavior.

IOP Conf. Series: Journal of Physics: Conf. Series **984** (2018) 012014 doi:10.1088/1742-6596/984/1/012014

1. Introduction

Hybrid organic-inorganic materials have been the challenge for the development of industry and the desire to cover new features in multiple domains offering the possibility of combining the dissimilar properties of organic and inorganic components [1-6]. The synthesis of new hybrid materials based on sulfate, phosphate and arsenate may contain original physical properties, is one of several research studies in chemistry laboratories in the world, due to its importance in both technological and industrial applications [1-5,7-9]. The cohesion forces in these hybrid compounds are dominated by electrostatic interactions, Van der Waals contacts and hydrogen bonds [10-17]. In fact, our global target is the synthesis of sulfate compounds associated to a wide range of metals which is then followed by the determination of the crystal structure and the nature of the hydrogen bonds that are one of the keys to understand the physical behavior of these materials. In this context, we report, herein, the chemical preparation, the crystallographic description, the spectroscopic characterization and the thermal behavior of $(C_6N_2H_{16})[Co(H_2O)_6](SO_4)_2.2H_2O$ which possesses a layered structure with the tran-1,4-diaminocyclohexane (DACH) molecule acting as structure-directing agent.

2. Experimental

2.1. Materials

$CoSO_4.7H_2O$ (Prolabo), H_3PO_3 (Sigma Aldrich) and DACH (Sigma Aldrich) were acquired from commercial sources and used as received.

2.2. Synthesis and characterization

The cobalt sulfate $(C_6N_2H_{16})[Co(H_2O)_6](SO_4)_2.2H_2O$ was synthesized in the presence of DACH as the structure directing agent at 293 K. A mixture of cobalt sulfate hydrate, phosphorous acid and DACH was shaken in distilled water. The resulting pink solution with a pH of 1.5 was leaved at room temperature for 2 weeks. The harvested product, consisting of single crystals in the form of pink-orange hexagonal was recovered by filtration, washed with a small amount of distilled water before being dried in ambient air.

2.3. Infrared Spectroscopy and Thermal behavior

Infrared spectrum of the compound was recorded on a VERTEX 70 FTIR Spectrometer in the range 4000-400 cm^{-1} using the ATR technique at 4 cm^{-1} resolution. Thermogravimetric analysis (TGA) data were recorded on an SDT-Q600 analyzer from TA instruments. The temperature varied from RT to 900°C at a heating rate of 10°C min^{-1}. Measures were carried out on samples in open platinum crucibles under air flow.

2.4. Crystal structure determination

Single-crystal X-ray diffraction measurement was carried out at room temperature using an Agilent Gemini S diffractometer equipped with a CCD detector and Mo radiation source. Acquired data was processed with the CrysAlisPro software[18]. Using Olex2 [19], the structure was solved with the olex2.solve [20] structure solution program using Charge Flipping and refined with the olex2.refine [20] refinement package using Gauss-Newton minimisation. All non-hydrogen atoms were refined anisotropically and hydrogen atoms were included in the model at calculated positions, refined with a rigid model with their U_{iso} value to $1.2U_{eq}$ of their parent atoms.

Table 1. Summary of crystal data and structure refinement.

CCDC no.	1588710
Chemical formula	$(C_6N_2H_{16})[Co(H_2O)_6(SO_4)_2].2H_2O$
Formula weight	255.70
Crystal system	Monoclinic
Space group	$P\,2_1/n$
a, Å	6.2897(2)
b, Å	12.3716(6)
c, Å	13.1996(4)
β, deg	98.091(3)
Volume, Å^3	1016.89(7)
Z	4
ρ_{calc} g cm^{-3}	1.6700
F(000)	541.6
μ(Mo Kα), mm^{-1}	1.126
Reflns (collected / unique, (Rint))	3245 / 1958 / 0.0209
Data / restraints / params	1958 / 0 / 131
R_1, wR [$F^2 > 2\sigma(F^2)$]	0.0368 / 0.0923
Goodness-of-fit on F^2	1.010
largest diff peak and hole (e Å^{-3})	0.47 / -0.42

3. Results and discussion

3.1. Crystal structure of $[(C_6N_2H_{16})[Co(H_2O)_6](SO_4)_2.2H_2O]$

The synthesized hybrid material with the general formula $(C_6N_2H_{16})[Co(H_2O)_6](SO_4)_2.2H_2O$ crystallizes in the monoclinic symmetry, space group $P\,2_1/n$. As can be seen in (Figure 1), the building unit of the hybrid compound consists of 21 non-hydrogen atoms, of which one Co atom is crystallographically independent. The crystal structure exhibits a 3D framework made upon isolated $Co(H_2O)_6$ octahedra, regular SO_4 tetrahedra, protonated DACH and water molecules, which interact through an intricate network of H-Bonds. Figure 2 depicts a projection of the crystal structure onto [100]. The inorganic part is constructed by the metal octahedral, sulfate tetrahedral and the free water molecules which are interconnected and stabilized via O-H···hydrogen bonds only (Figure 3). The divalent cation Co^{2+} lies in a special position on an inversion center (Wyckoff site 2b), it is octahedrally coordinated by six O atoms that all belong to water molecules. The Co-O bond lengths range from 2.047(3) to 2.113(3) Å with an average Co-O distance of 2.076 Å which is comparable to that of 2.072 Å reported for $Co_4(SO_4)(OH)_6(C_2N_2H_8)_{0.5}.3H_2O$ [21] and to that of 2.092 Å for $(DACH)_{0.5}Co_3(H_2O)(OH)(PO_4)(HPO_4)$ $(3+x)H_2O$ [22]. Within the sulfate group, The lengths S-O bonds exhibit a mean value of 1.466 Å which is very close to the corresponding distance of 1.468(4) Å obtained from $La_2(H_2O)_2(C_2H_{10}N_2)_3(SO_4)_4$ [23] and 1.474 Å in $Co(SO_4).6D_2O$ [24]. The DACH dications adopts the chair conformation and are placed in the (100) plane with the C-NH_3^+ groups along the *a*-direction of the unit cell (Figure 4). The diprotonated organic molecule resides in the interlayer region and interacts with the host network through extensive hydrogen bonds. In another view, the structure might be described as consisting of strictly alternating CoO_6 octahedra and SO_4 tetrahedra forming an eight membered ring via hydrogen bonds between the $O_{(5)}$, $O_{(7)}$ and $O_{(11)}$ from Co atom and the $O_{(4)}$, $O_{(8)}$ atoms of the SO_4 groups. The free water molecules are also located within eight membered ring apertures and participate in hydrogen bonding with the framework contributing to the structural stability of the compound. The diprotonated DACH molecules are inserted into the inorganic layers and interact with the mineral part through connecting the sulfate groups by N-H···O hydrogen bonds (Figure 5). The hydrogen bonds (D-H···A) topology found in $(C_6N_2H_{16})[Co(H_2O)_6](SO_4)_2.2H_2O$ gives rise to a three dimensional network that certainly contributes to increasing its stability.

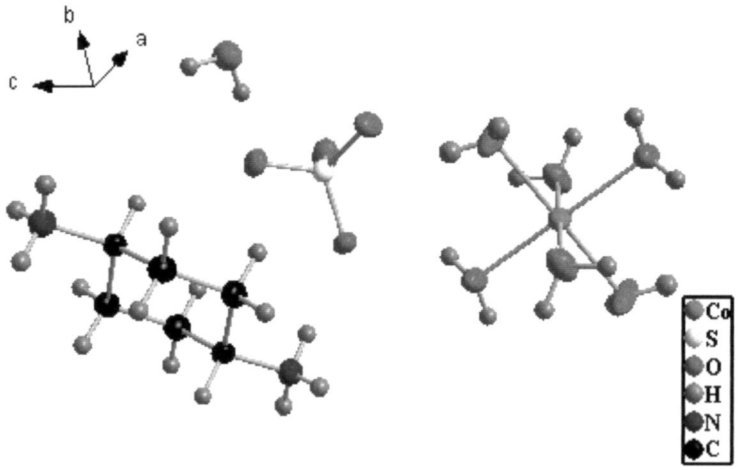

Figure 1. Building unit in $(C_6N_2H_{16})[Co(H_2O)_6](SO_4)_2.2H_2O$, ellipsoids are shown at 50% probability.

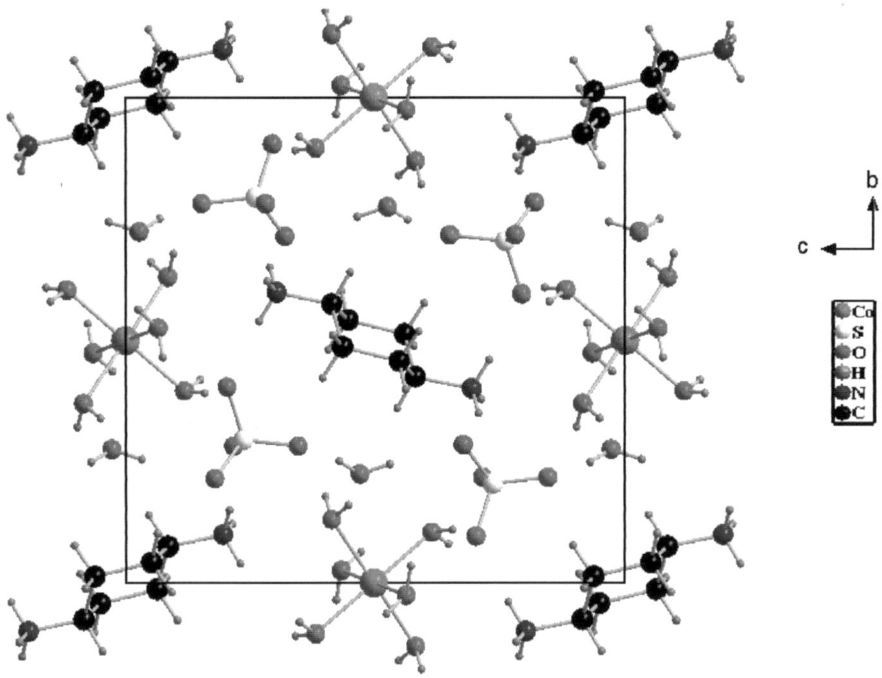

Figure 2. Projection of the crystal structure of $(C_6N_2H_{16})[Co(H_2O)_6](SO_4)_2.2H_2O$ on (100).

IOP Conf. Series: Journal of Physics: Conf. Series **984** (2018) 012014 doi:10.1088/1742-6596/984/1/012014

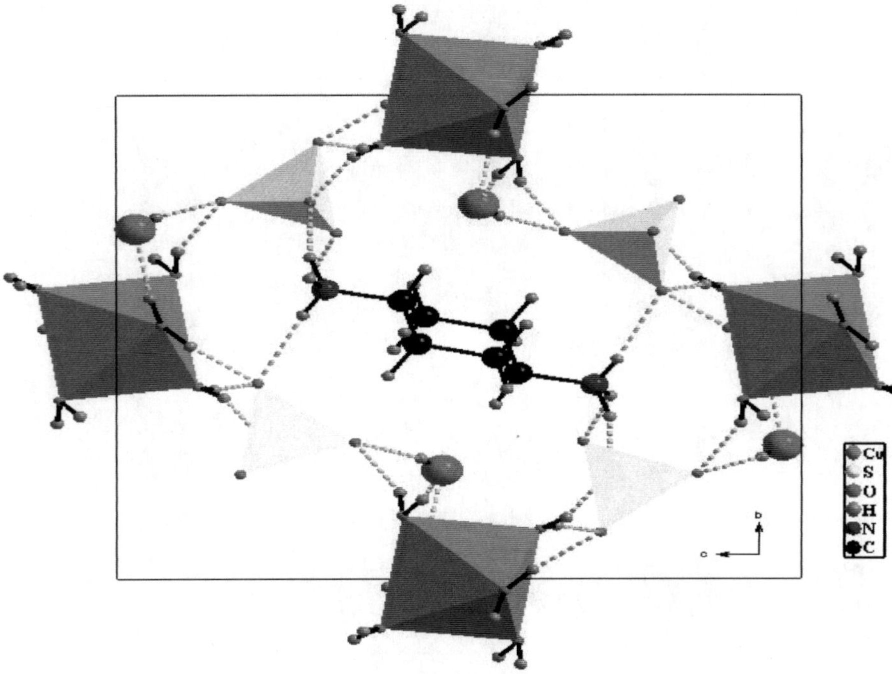

Figure 3. View of crystal structure $(C_6N_2H_{16})[Co(H_2O)_6](SO_4)_2.2H_2O$ along the *a* axis.

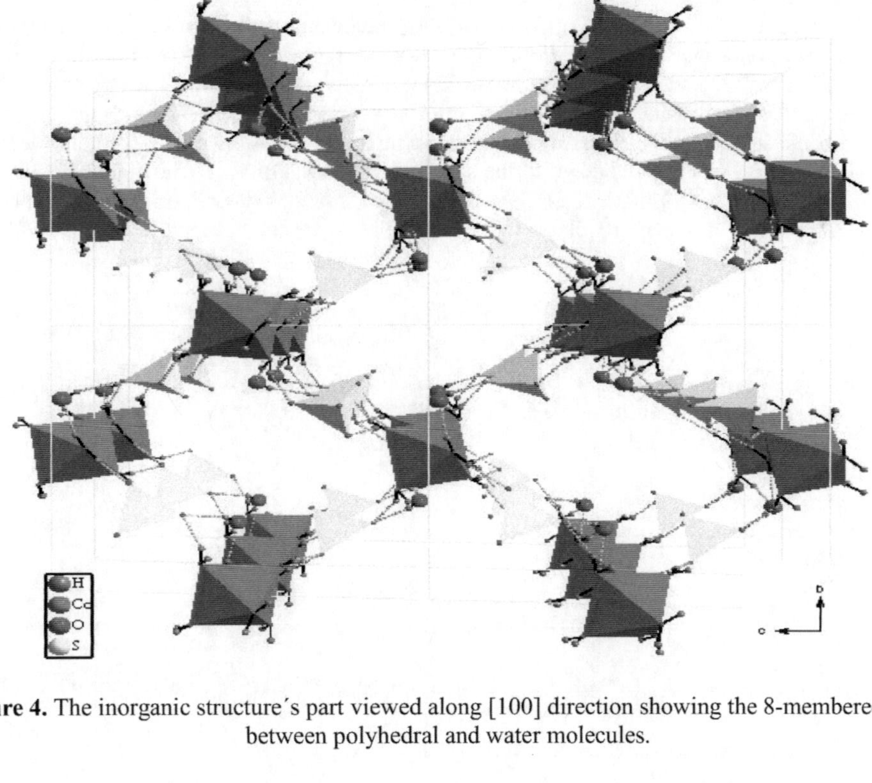

Figure 4. The inorganic structure´s part viewed along [100] direction showing the 8-membered rings between polyhedral and water molecules.

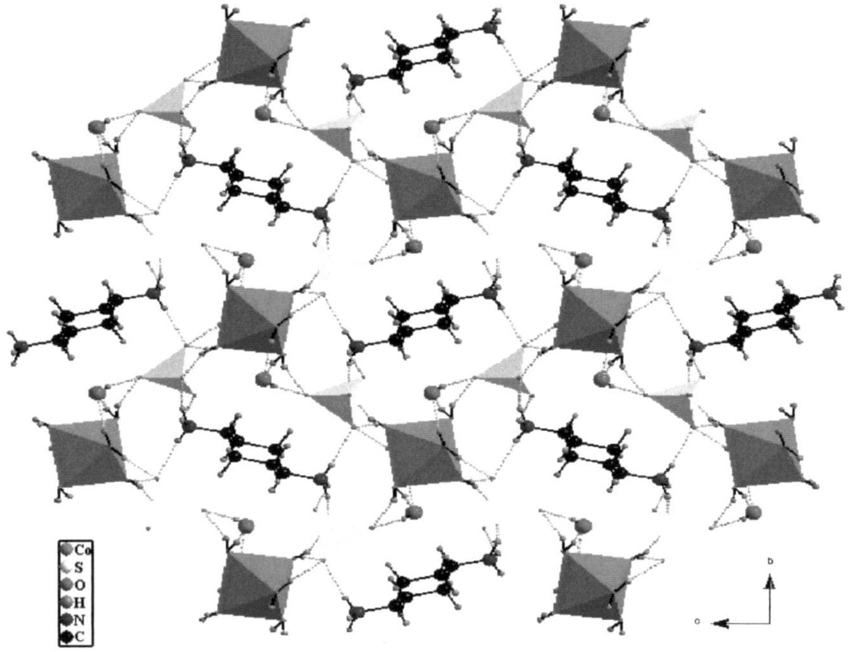

Figure 5. Projection of the crystal structure along *a*. Dashed green lines indicate hydrogen bonds.

3.2. Infrared spectroscopy

The infrared spectrum of $(C_6N_2H_{16})(OH)[Co(H_2O)_6]SO_4$ phase exhibits the bands corresponding to the vibrations of the DACH and the sulfate group. As shown in Figure 6 the stretching vibration of NH_2 was in 3098-3213 cm^{-1} region. The absorption at 1034 cm^{-1} is due to the symmetric stretching of SO_4 groups, whereas the small band centered at 592 cm^{-1} is attributed to the bending vibration of SO_4 region. The band at 1647 cm^{-1} corresponds to the symmetric vibration of water molecules. The vibration at around 2919 cm^{-1} boils down to the stretching vibration of C-H while band at 1399 cm^{-1} corresponds to the stretching vibrations of C-N, the value at around1490 cm^{-1} belongs to the bending vibration of CH_2, these vibrations are related to the organic template cation [25,26]

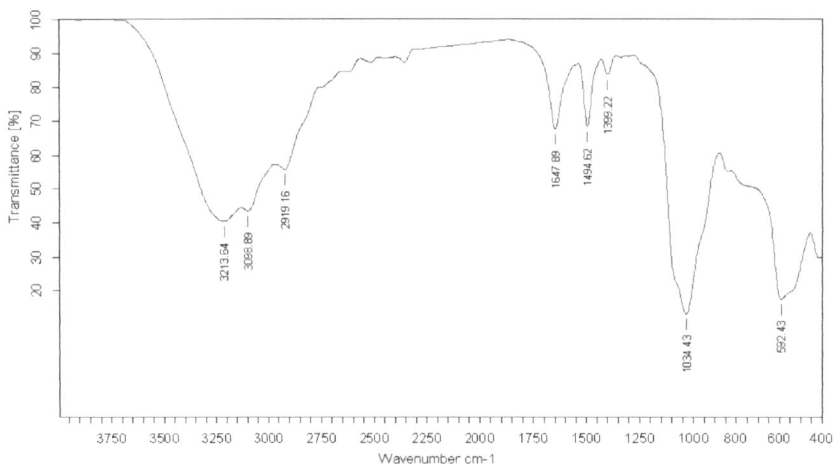

Figure 6. IR spectrum of $(C_6N_2H_{16})[Co(H_2O)_6](SO_4)_2.2H_2O$.

3.3 Thermal behavior

Data given from thermogravimetric analysis (Figure 7) shows three cleanly separated steps of weight loss. $[(C_6N_2H_{16})[Co(H_2O)_6](SO_4)_2.2H_2O]$ undergoes a first mass loss of 31.5% between room temperature and 200°C accompanied by an endothermic DSC signal. The sharp weight coincidences with the degradation of organic moiety, together with the departure of 2.5 water molecules (calculated value 31.5%). The second stage (theoretical weight loss 13.49%) starting at about 250°C and ending at 550°C is in agreement with the removal of 3.5 water molecules(calculated mass loss 12.33%), this phenomenon is coupled with an endothermic and an exothermic heat flow observed at about 300°C and 400°C respectively. The slower departure of the remaining water molecules together with the partial release of sulfate groups is shown between 650°and 850°C which is attended with an endothermic curve on the DSC curve and leads to a final crystalline compound (observed weight loss: 24.67% calculated weight loss: 25.8%).

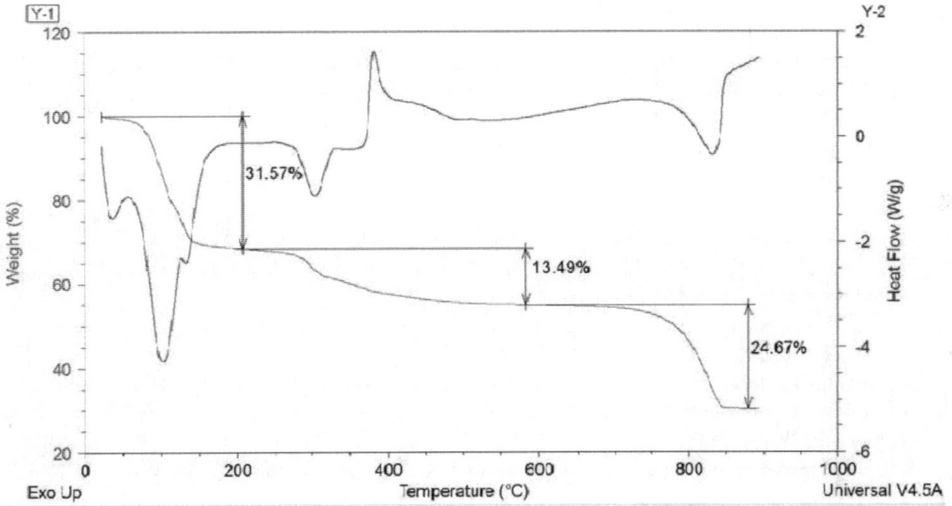

Figure 7. DSC-TG curves for the decomposition of the $(C_6N_2H_{16})[Co(H_2O)_6](SO_4)_2.2H_2O$.

4. Conclusion

A new cobalt sulfate templated by the trans-1,4-diaminocyclohexane molecule $(C_6N_2H_{16})[Co(H_2O)_6](SO_4)_2.2H_2O$ was prepared by slow evaporation method. The obtained crystals were investigated by FTIR spectroscopy and crystallographically characterized by single-crystal X-ray diffraction. The crystal structure consists of isolated ions $(SO_4)^{2-}$, $[C_6N_2H_{16}]^{2+}$ and $[Co(H_2O)]^{2+}$. These different entities are linked and stabilized via two types of hydrogen bonds N-H···O and O-H···O. The infrared spectrum shows bands characterizing the sulfate group, tran-1,4-diaminocyclohexane (DACH) and water molecules. The thermogravimetric analysis showed that the dehydration of $(C_6N_2H_{16})[Co(H_2O)_6](SO_4)_2.2H_2O$ takes place in three steps, the thermal decomposition of the phase consists mainly in the loss of the organic moiety and one sulfate group leading to the formation of cobalt sulfate.

Acknowledgments

BEB would thank Prof. Aurelio Cabeza Díaz, Departamento de Química Inorgánica, Cristalografía y Mineralogía, Facultad de Ciencias, Universidad de Málaga (Spain) for the thermal measurements.

References

[1] Sanchez C, Belleville P,Popall M and Nicole L 2011 Chem. Soc. Rev. **40** 696
[2] Mohana Reddy A. L, Gowda S. R, Shaijumon M. M and Ajayan P. M 2012 *Adv.Mater.***24**, 5045
[3] Sanchez C, Julián B, Belleville P, Popall M and Nicole L 2011 *Chem. Soc. Rev* **40** 696
[4] Dolbecq A, Dumas E, Mayer C. R and Mialane P, 2010 *Chem. Rev* **110** 6009
[5] Rao C. N. R, Cheetham A K and Thirumurugan A J 2008 *Phys. Condens. Matter* **20** 083202
[6] Qian L, Willneff E and Zhang H 2009 *Chem. Commun* 3946
[7] Leblanc N, Allain M, Mercier N and Sanguinet L 2011 *Cryst. Growth Des* **11** 2064
[8] Bi W, Louvain N, Mercier N, Luc J, Rau I, Kajzar F and Sahraoui B 2008 *Adv. Mater* **20** 1013
[9] Wight A. P and Davis M.E 2002 *Chem. Rev* **102** 3589
[10] Gesi K 1976 *Phys. Status Solidi A* **33** 479
[11] Schwalowsky L,Vinnichenko V, Baranov A, Bismayer U, Merinov B and Eckold G 1998 *J. Phys.Condens.Matter* **10** 3019
[12] Gesi K 1977 *J. Phys. Soc.* Jpn **42** 1785
[13] Sooryanarayana K and Guru Row T. N 1996 *Phase Transitions* **58** 263
[14] Fukami T, Horiuchi K, Nakasone K and Furukawa K 1996 *Jpn. J. Appl. Phys* **35** 2253
[15] Suzuki S and Makita Y 1978 *Acta Crystallogr. B* **B34** 732
[16] Kamoun M, Ben Ghozlen M H and Daoud A 1987 *Phase Transitions* **9** 247
[17] Chen R H, Wang Li-Ming and Yang S C 1992 *phase Transitions* **37** 141
[18] CrysAliPro Software System,version 1.171.38.41, Rigaku Corporation: Oxford,UK
[19] Dolomanov O, Bourhis L, Gildea R, Howard J and Puschmann H 2009 *J. Appl. Cryst.* **42** 339
[20] Bourhis L, Dolomanov O, Gildea R, Howard J and Puschmann H 2015 *Acta. Cryst.***A71** 59
[21] Rujiwatra A, Kepert C J, Claridge J B, Rosseinsky M J, Kumagai H and Kurmoo M 2001 *J.Am.Chem.Soc* **123** 10584
[22] Chang W K, Chiang R K and Wang S L 2007 *J.Solid state chem* **180** 1713
[23] Bataille T and Louër D 2002 *J.Mater.Chem* **12** 3487
[24] Kellersohn T, Delaplane R G, Olovsson I and Mclntyre G J 1993 *Acta Crystallogr* **B49** 179
[25] Nakamoto K 1997 *Infrared and Raman Spectra of Inorganic and Coordination Compounds* (NewYork: Wiley)
[26] Stuart B 2004 *Infrared Spectroscopy: Fundamentals and Applications* (John Wiley & Sons, Inc)

$Cs_2[M(H_2O)_6]_3(HPO_3)_4$, M= Co, Ni: Crystal structures, IR and thermal studies

S Chaouch[1], R Ouarsal[1], M Akouibaa[1], S Rakib[1], M Lachkar[1], B El Bali[1], M Dusek[2]

[1]Engineering Laboratory of Organometallic and Molecular Materials, Faculty of Sciences, University Sidi Mohamed Ben Abdellah, Po. Box 1796 (Atlas), 30000 Fez, Morocco.
[2]Institute of Physics ASCR, v.v.i., Na Slovance 2, 182 21 Praha 8, Czech Republic.

Email: chaouch2005@yahoo.fr

Abstract. $Cs_2[M(H_2O)_6]_3(HPO_3)_4$, M= Co (**1**) Ni (**2**) are two isostructural phosphites. They crystallize in monoclinic (S.G.: C2/m, Z = 2) and the following cell parameters (Å, °): [a = 12.6132(5), b = 7.2488(3), c = 16.7990(6), β = 92.591(4), V = 1534.37(10) Å^3 for (1) and a = 12.5509(3), b = 7.2452(2), c = 16.6690(4), β = 92.490(2), V = 1514.34(7) Å^3 for (2)]. The structure consists of isolated layers parallel to (001) crystallographic plane made from corner sharing $[M(H_2O)_6]$ (M= Co, Ni) and $[Cs(H_2O)_8]$ polyhedra. The HPO_3 phosphite ions are linked to the $Cs(H_2O)_8$ and $M(H_2O)_6$ polyhedra via hydrogen bonds. The 3D framework is build upon these layers, which are interconnected by an intricate network of hydrogen bonds. IR spectra of the title compounds present the HPO_3 characterizing bands. TG curves document the process of the full dehydration.

Keywords: Cs-phosphites, Crystal Structure, X-Rays, IR, TGA, TDA.

1. Introduction

Materials science has been growing very quickly during the last decades, giving rise to numerous new compounds with a great diversity in compositions and features. Beyonds these, open-framework metal phosphites have been intensively investigated. Much attention has been paid to transitional metal phosphites due to their rich structural chemistry and wide potential applications in ion-exchange, adsorption, separation and catalysis [1-6]. Since the properties of inorganic materials can be tuned by their size and shape, the preparation of transition metal phosphites nanostructures with certain morphologies and sizes has been attracting increasing research interest [7, 8]. Phosphite HPO_3 group has been investigated as a possible replacement for the traditional phosphate tetrahedron with great success paving the way for the formation of a new family of compounds [9-13]. These compounds may be simple, mixed, inorganic or hybrid organic-inorganic. We cite for example $Li_2HPO_3.H_2O$ [14], $Pb_2(HPO_3)(NO_3)_2$ [15], $NaMg(H_2PO_3)_3.H_2O$ [16], $Fe_2(HPO_3)F_2$ [17], $(C_6H_5NH_3)ZnCl(HPO_3)$ [18], and $K_2[Mn_3](HPO_3)_4]$ [19]. $[HPO_3]^{2-}$ is pseudo-tetrahedron but with only three vertices that can be available for sharing by other polyhedra, unlike four in PO_4 tetrahedron, these features is expected to give rise to the structures differing from four-connected polyhedron compounds [20]. In 2011, Zhou [21] reported on Proton-conducting solid acid electrolytes based upon $MH(HPO_3)$, M = Alkaline or NH_4^+.

In this study, we report on the synthesis and crystal structures of two new mixed phosphites: Disaesium tri [hexahydrate M(II)] tetrakis(hydrogenphosphite $Cs_2[M(H_2O)_6]_3(HPO_3)_4$ (M= Co, Ni). Their IR spectra and their thermal features are discussed in correlation with their structural data.

2. Experimental

2.1. Synthesis

To synthesize $Cs_2[M(H_2O)_6]_3(HPO_3)_4$ [M= Co (**1**) and Ni (**2**)], two solutions were prepared separately: (A) a solution of Cs_2CO_3 (5.8 mmol) is added to 5.8 mmol of H_3PO_3 until pH 8 is reached. (B) 10 mL of H_3PO_3 (1M) were prepared. Solution (A) is added dropwise to solution (B) followed by the addition of 5 mmol of $MCl_2.6H_2O$ (M= Co, Ni). The mixture is stirred during 2h at 333K. A precipitate is obtained, which is filtered out and the filtrate allowed standing at room temperature. After two weeks, respectively pink and light green colored crystals for (**1**) and (**2**) were collected and washed with an (80:20 v/v) water-ethanol mixture.

2.2. Infrared Spectroscopy and Thermal Analysis

Infrared spectra of the compound were recorded on a VERTEX 70 FTIR Spectrometer in the range 4000-400 cm-1 using the ATR technique.

2.3. Crystal structure refinements

The X-ray diffraction data were collected on an Oxford Diffraction four-circles X-ray diffractometer XCALIBUR using graphite monochromatized MoKα radiation (λ= 0.71073 Å) with Mo-Enhance collimator and Atlas CCD as a detector. The intensity data were corrected for Lorentz and polarization effects. A numeric analytical absorption correction was carried out with the program CrysAlis RED [22]. The structures were solved with the Direct Methods procedure of the SUPERFLIP program [23] and refined by a full-matrix least-squares technique based on F^2 using Jana2006 [24]. All non-hydrogen atoms were refined with anisotropic displacement parameters. Positions of hydrogen atoms belonging to water molecules and to phosphorus were found in the difference Fourier map and refined with a restraint on their distance to the parent oxygen atom. Crystal and refinement data for the two new compounds are reported in Table 1. The isotropic temperature parameters of hydrogen atoms were fixed as 1.2*Ueq of the parent atom. Supplementary tables of crystal structures and refinements have been deposited [25].

3. Results and discussion

3.1. Crystal structure description

$Cs_2[M(H_2O)_6]_3(HPO_3)_4$, M= Co (1) and Ni (2), are two isostructural new mixed phosphites, both crystallizing with monoclinic symmetry (S. G.: C2/m). As shown in Figure 1, the building unit of both compounds contains two crystallographic distinct metal-M, two P and one Cs atoms. The crystal structures of these compounds can be described as a three-dimensional network made up of $[CoO_6]$ octahedra, $[CsO_8]$ polyhedra and $[HPO_3]$ tetrahedral (Figure 2). Divalent M^{2+} ions (M = Co and Ni) are octahedraly coordinated by six oxygen atoms from water molecules. The shape of all coordination polyhedra is partially controlled by symmetry because M1 is located in a centre of symmetry (site 2/m) while M2, P1, P2 and Cs lie in a mirror plane. Cs atom is coordinated by eight oxygen atoms from eigh different water molecules. $Cs(H_2O)_8$ polyhedra is irregular and share all their corners with neighboring $M(H_2O)_6$ octahedra.

Table 1. Crystal data and details of the two isostructural (Monoclinic, C2/m) $Cs_2[M(H_2O)_6]_3(HPO_3)_4$, M = Co (**1**), Ni (**2**).

Chemical formula	**$Cs_2[Co(H_2O)_6]_3(HPO_3)_4$**	**$Cs_2[Ni(H_2O)_6]_3(HPO_3)_4$**
Crystal system	Monoclinic	Monoclinic
Space group	C2/m	C2/m
Molecular weight (g /mol)	1086.8	1086.1
Radiation (Kα Mo) Å	0.71073	0.71073
Unit-cell dimensions	a = 12.6132(5) Å b = 7.2488(3) Å c = 16.7990(6) Å β = 92.591(4)°	a = 12.5509(3) Å b = 7.2452(2) Å c = 16.6690(4) Å β = 92.490(2)°
Volume Å^3	1534.37(10) Å^3	1514.34(7) Å^3
Z	2	2
Dcal (g/cm^3)	2.3516	2.3811
Absorption coefficient (mm^{-1})	4.238	4.499
F(000)	1062	1068
Crystal size (mm)	0.2290× 0.2078 × 0.1572	0.3809 × 0.3432 × 0.2298
Diffractometer	Oxford Diffraction Xcalibur	Oxford Diffraction Xcalibur
Range θ: θmin- θmax (°)	2.8034-26.4922	3.24-26.5
Limiting index	h = -15→15 k = -8→9 l = −21→21	h = -15→15 k = -9→9 l = −20→20
Reflns (collected/ unique)	9960 / 1721	9552 / 1698
Reflns with $I > 3\sigma(I)$	1269	1534
R factors ($I > 3\sigma(I)$)	R_1 = 0.0373 WR_2 = 0.0927	R_1 = 0.0256 WR_2 = 0.0821
S(F²)	1.69	1.69
$(\Delta/\sigma)_{max}$	0.0094	0.0403
$(\Delta\rho)_{max}$; $(\Delta\rho)_{min}$ (e.Å^{-3})	1.53 ; -0.66	0.64 ; -0.33

$M(1)(H_2O)_6$ octahedra is connected to two neighboring $Cs(H_2O)_8$ polyhedra by sharing two adjacent edges with each one of them, however $M(2)(H_2O)_6$ shares five of its vertices with three various $Cs(H_2O)_8$ polyhedra, leading to layers extended in the ab plan (Figure 3 a&b). HPO_3 are linked to the polyhedrons $Cs(H_2O_8)$ and $M(H_2O)_6$ only through hydrogen bonds. The connection between the neighbouring layers is assured by hydrogen bonds while The H atom belonging to phosphorus group is not involved in any hydrogen bonds. Lists of such bonds in (**1**) and (**2**) are summarized in Table 2.

Average $\overline{d}$ (Co-O) bond lengths is 2.080 Å, of the same magnitude as the values reported in similar compounds: $K_2Co(HPO_3)_2.2H_2O$ (2.105 Å) [26], $(NH_4)_2[Co(H_2O)_6]_3(HPO_3)_4$ (2.093 Å) [27] and $NaCo(H_2PO_3)_3.H_2O$ (2.101 Å) [28]. Average $\overline{d}$ (Ni-O) bond lengths is 2.050 Å, which can be compared to 2.070 Å in $NaNi(H_2PO_3)_3.H_2O$ [29], 2.057 Å in $(NH_4)_2[Ni(H_2O)_6]_3(HPO_3)_4$ [30] and 2.078 Å in $Na_2Ni(HPO_3)_2$ [31].

The $[MO_6]$ octahedra are isolated and connected only via the $Cs(H_2O)_8$ polyhedra. Short distances M-M are 6.678(13) Å and 6.636(6) Å respectively in (**1**) and (**2**). These values are a bit longer than the values reported in the literature: in $NaCo(H_2PO_3)_3.H_2O$ (5.061(3) Å) [28], in $(NH_4)_2[Co(H_2O)_6]_3(HPO_3)_4$ (6.1873(3) Å) [27], in $K_2Co(HPO_3)_2.2H_2O$ (5.319(2) Å) [26] and in $NaNi(H_2PO_3)_3.H_2O$ (5.4091(4) Å) [29]. This might be related as a first approach to the fact that Co and Ni in the present phosphites are both coordinated by six O atoms from water molecules, which are engaged in H-Bonds interactions. In the known phosphites like $NaCo(H_2PO_3)_3.H_2O$ [28] or $NaNi(H_2PO_3)_3.H_2O$ [29], they are rather coordinated by O atoms belonging to phosphites groups.

Table 2. Significant H-bonds in $Cs_2[M(H_2O)_6]_3(HPO_3)_4$, M= Co (**1**) Ni (**2**).

	D-H...A	D-H (Å)	H...A (Å)	D...A (Å)	Ang D-H...A (°)
$Cs_2[Co(H_2O)_6]_3(HPO_3)_4$	O(5)-H(5a)...O(2)	0.815 (17)	1.907 (18)	2.682 (4)	158.52 (141)
	O(5)-H(5a)...O(2)	0.818 (37)	1.868 (38)	2.685 (4)	175.84 (330)
	O(6)-H(6a)...O(2)	0.814 (37)	1.894 (37)	2.680 (3)	161.94 (366)
	O(7)-H(7a)...O(3)	0.812 (17)	1.921 (18)	2.725 (5)	170.19 (146)
	O(7)-H(7b)...O(3)	0.816 (25)	1.934 (24)	2.747 (5)	174.12 (194)
	O(8)-H(8a)...O(3)	0.820 (4)	1.914 (38)	2.703 (4)	161.40 (37)
	O(9)-H(9a)...O(4)	0.815 (51)	1.884 (51)	2.686 (6)	168.08 (509)
	O(9)-H(9b)...O(2)	0.821 (18)	1.849 (17)	2.664 (4)	170.99 (154)
	O(10)-H(10a)...O(1)	0.818 (18)	1.852 (18)	2.670 (6)	179.31 (158)
	O(10)-H(10b)...O(4)	0.813 (48)	1.876 (44)	2.685 (8)	173.57 (410)
$Cs_2[Ni(H_2O)_6]_3(HPO_3)_4$	O(5)-H(5a)...O(2)	0.818 (10)	1.875 (12)	2.678 (3)	166.88 (103)
	O(5)-H(5b)...O(1)	0.814 (26)	1.878 (28)	2.682 (3)	169.44 (234)
	O(6)-H(6a)...O(2)	0.810 (23)	1.889 (23)	2.678 (2)	164.76 (226)
	O(7)-H(7a)...O(3)	0.818 (14)	1.923 (14)	2.735 (3)	171.38 (130)
	O(7)-H(7b)...O(3)	0.816 (24)	1.946 (20)	2.746 (3)	166.37 (191)
	O(8)-H(8a)...O(3)	0.824 (28)	1.894 (26)	2.711(3)	170.70 (237)
	O(9)-H(9a)...O(4)	0.821 (32)	1.871 (32)	2.691 (3)	177.55 (312)
	O(9)-H(9b)...O(2)	0.820 (32)	1.861 (13)	2.673 (3)	170.48 (118)
	O(10)-H(10a)...O(1)	0.818 (21)	1.875 (17)	2.687 (4)	171.64 (170)
	O(10)-H(10b)...O(4)	0.821 (30)	1.869 (28)	2.686 (5)	173.38 (268)

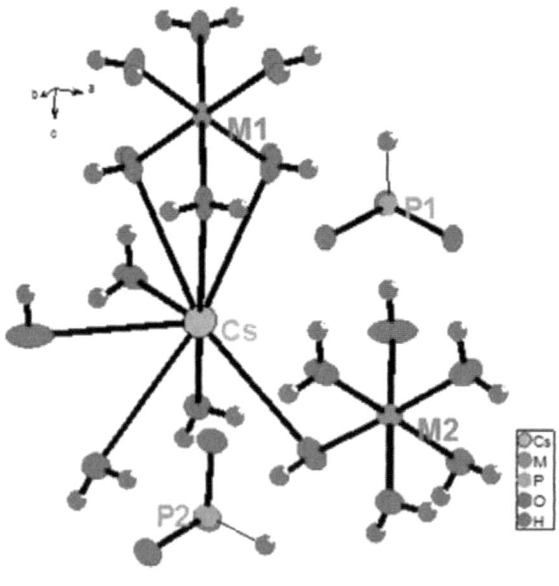

Figure 1. Building unit of $Cs_2[M(H_2O)_6]_3(HPO_3)_4$ [M = Co and Ni], shown with 50% probability displacement ellipsoids.

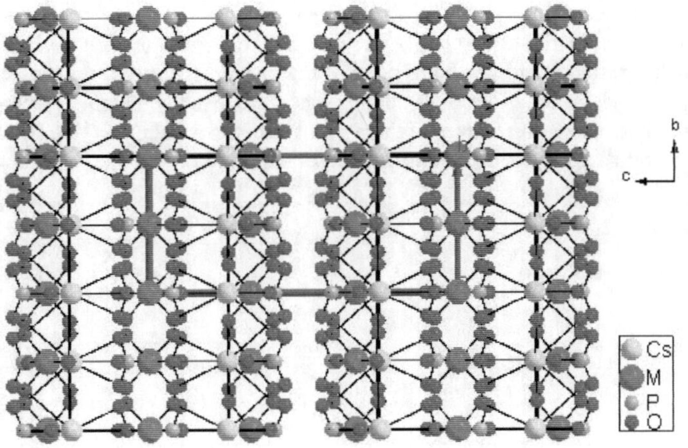

Figure 2. A view of the structure of $Cs_2[M(H_2O)_6]_3(HPO_3)_4$ along the a axis (All H atoms are omitted for clarity).

Cs-O bond distances range from 3.340(4) Å to 3.566(5) Å with averages $\overline{d}$ (Cs-O) distances of 3.456 Å and 3.434 Å respectively in (**1**) and (**2**). These values are comparable with those reported in $CsIn(HPO_3)_2$, 3.411 Å [32], α-$CsH_2PO_3.H_3PO_3$, 3.217 Å [33] and β-$CsH_2PO_3.H_3PO_3$, 3.310 Å [33].

P atoms are bound with one H atom and three O atoms. The averages $\overline{d}$ (P-H) distances are 1.397 Å and 1.322 Å respectively in (**1**) and (**2**) whilst the mean P-O bond length is 1.525 Å in both compounds. These values are generally similar with the values found in the literature [26, 27, 30, 34].

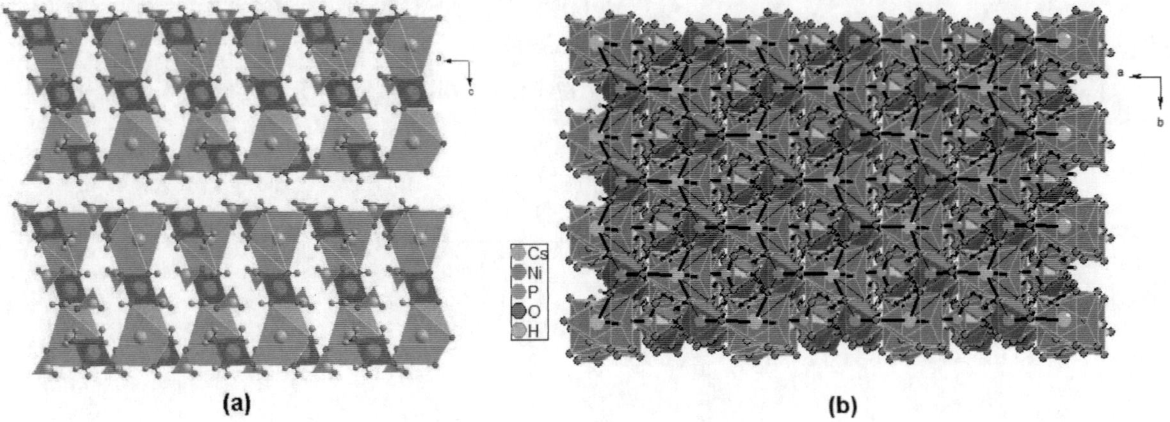

Figure 3. Polyhedral representation of the crystal structure along the b-axis **(a)** and c-axis **(b)**.

3.2. Infrared measurement.

The infrared spectra of the two new mixed phosphites are undistinguishable; as representative spectra we report on Figure 4 (a&b) respectively that of $Cs_2[M(H_2O)_6]_3(HPO_3)_4$ and $Cs_2[Ni(H_2O)_6]_3(HPO_3)_4$. Frequencies of vibration are summarized in Table 3. The IR spectra show the symmetric and anti-symmetric stretching vibrations of PO_3 group bands at 655 and 1042 cm^{-1}. The presence of bands around 564 and 471 cm^{-1} characterize symmetric and anti-symmetric P-O deformation modes. The sharp peak at 2423 cm^{-1} is attributed to the stretching vibration of the P-H bond while the bending vibrations δ(P-H) appear around 900 cm^{-1}. The bands corresponding to stretching vibration and deformation of the OH group of the water molecule are observed about 3000 and 1560 cm^{-1} as previously reported in the literature [35, 36]. In fact, in the study relating the frequency of vibration of the O-H bond to the distance O...O, it was concluded that Hydrogen bonding causes a displacement of the vibration towards the low frequency. Thus υ(OH) is of the order of 3000 cm^{-1} when the distance Donor...Acceptor is of the order of 2.68 Å [35, 36]. One interesting feature of the present structures is the fact that HPO_3 phosphite group is practically free compared to its behaviour in above-mentioned known structures where it generally

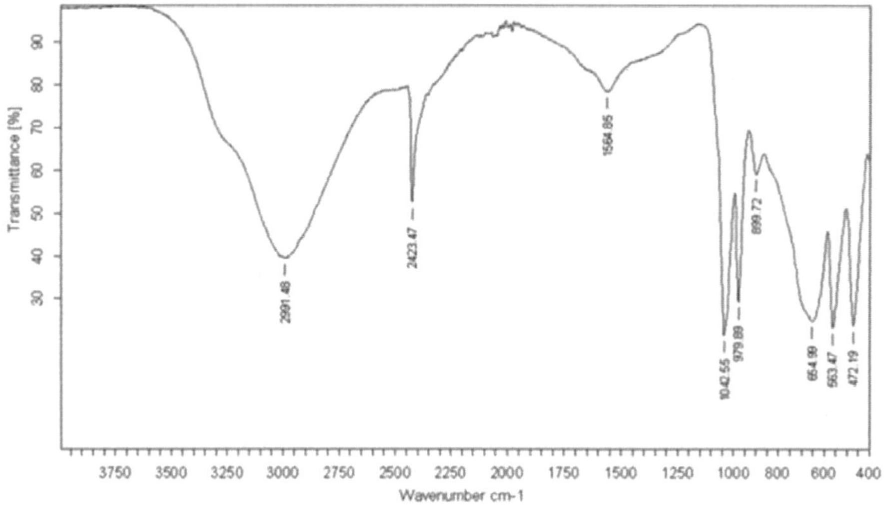

shares corners with the metal polyhedra. Mobility freedom of HPO_3 in the framework of the new compounds results in lower resolution in their IR spectra.

(a)

(b)

Figure 4. The IR spectra of **(a)** $Cs_2[M(H_2O)_6]_3(HPO_3)_4$ and **(b)** $Cs_2[Ni(H_2O)_6]_3(HPO_3)_4$, with band assignments.

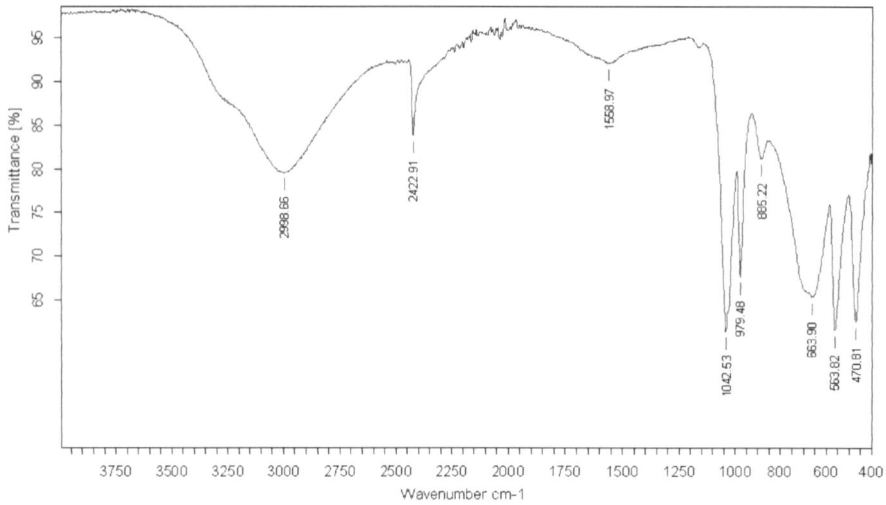

Table 3. IR bands assignments for $Cs_2[M(H_2O)_6]_3(HPO_3)_4$, M= Co (**1**) Ni (**2**).

Attribution	$Cs_2[Co(H_2O)_6]_3(HPO_3)_4$	$Cs_2[Ni(H_2O)_6]_3(HPO_3)_4$
υ(OH)eau	2998	2991
υ(PH)	2423	2423
δ(OH)eau	1559	1565
$υ_{as}(PO_3)$	1042	1042
δ(PH)	979	978
	885	900
$υ_s(PO_3)$	664	655
$δ_s(PO_3)$	564	563
$δ_{as}(PO_3)$	471	472

3.3. Thermal TGA and DTA analysis

As in the IR study, TGA and DTA curves of the title compounds are also almost the same. Figure 5 depicts the results from $Cs_2[Ni(H_2O)_6]_3(HPO_3)_4$ as representative case. Analysis shows a continuous weight loss of 28.92 % for the title compounds in the range of 50-200 °C. This is obviously attributed to the loss of 18 water molecules, and it is in agreement with the calculated amount of the total removal of water (29.8%) for both compounds. On the TDA study, this appears as an exothermic peak at 112 °C.

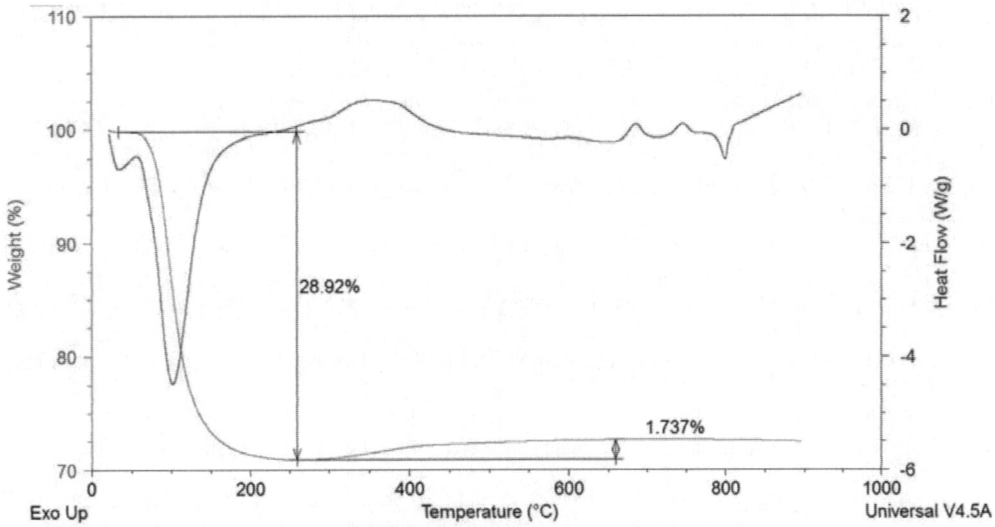

Figure 5. TG-TDA curves of $Cs_2[Ni(H_2O)_6]_3(HPO_3)_4$

Conclusion

Two new Caesium-metal phosphites with the formula $Cs_2[Co(H_2O)_6]_3(HPO_3)_4$ and $Cs_2[Ni(H_2O)_6]_3(HPO_3)_4$ have been synthesized and structurally characterized. The crystal lattice can be described as layers formed of MO6 octahedra linked via CsO8 polyhedra. $[HPO_3]^{2-}$ are linked to $Cs(H_2O)_8$ and $M(H_2O)_6$ via hydrogen bonds. IR spectra of the two new phosphites present the bands characterising the group HPO_3. TGA confirm a full loss of the crystalline water during the heating process. The phenomenon is also confirmed by an exothermic peak in the TDA curves around 112°C.

Acknowledgements

The authors acknowledge the financial support from CNRST (Morocco) (URAC 19), CURI-Fez and the grant "Praemium Academiae" of the Academy of Sciences of the Czech Republic.

References

[1] Xiong D B, Li M R, Liu W, Chen H H, Yang X X and Zhao J T 2006 *J. Solid State Chem.* **179** 2571-77
[2] Wang L, Song TY, Fan Y, Tian Z F, Wang Y, Shi S H and Xu J N 2006 *J. Solid State Chem.* **179** 3400-05
[3] Chung U C, Mesa J L, Pizarroa J L, Jubera V, Lezamab L, Arriortua M I and Rojo T 2005 *J. Solid State Chem.* **178** 2913-21
[4] Shi S H, Qian W, Li G H, Wang L, Yuan H M, Xu J N, Zhu G S, Song T Y and Qiu S L 2004 *J. Solid State Chem.* **177** 3038-44
[5] Lin Z, Fan W, Gao F, Chino N, Yokoi T and Okubo T 2006 *J. Solid State Chem.* **179** 723-28
[6] Zhong Y J, Chen Y M, Sun Y Q and Yang G Y 2005 *J. Solid State Chem.* **178** 2613-19
[7] Whitesides G M and Grzybowski B A 2002 *Science* **295** 2418-21
[8] Park S, Lim J H, Chung S W and Mirkin C A 2004 *Science* **303** 348-51
[9] Rodgers J A and Harrison W T A 2000, *Chem. Commun.* **23** 2385-86
[10] Liang J, Li J Y, Yu J H, Chen P, Fang Q R, Sun FX and Xu R R 2006 *Angew. Chem. Int. Ed.* **45** 2546-48
[11] Fernandez S, Mesa J L, Pizarro J L, Lezama L, Arriortua M I and Rojo T 2002 *Angew. Chem. Int. Ed.* **41** 3683-85
[12] Chung U C, Mesa J L, Pizarro J L, Lezama L, Garitaonandia J S, Chapman J P and Arriortua M I 2004 *J. Solid State Chem.* **177** 2705-13
[13] Yang Y L, Li N, Song H B, Wang H G, Chen W B and Xiang S H 2007 *Chem. Mater.* **19** 1889-91
[14] Chaouch S, Ouarsal R, El Bali B, Lachkar M, Bolte M and Dusek M 2010 *J. Chem. Crystallogr.* **40** 526-30
[15] Ouarsal R, Lachkar M, Dusek M, Fejfarova K and El Bali B 2009 *Acta Crystallogr.* **E65** i35-37
[16] Ouarsal R, Alaoui A T, Lachkar M, Dusek M, Fejfarova K and El Bali B 2003 *Acta Crystallogr.* **E59** i33-i35
[17] Liu L, Wang X, Xu L, Liu X, Liu L, Bi B, and Pang W 2009 *Inorg. Chim. Acta* **362** 3881-84
[18] Chaouch S, Ouarsal R, Lachkar M, Capitelli F and El Bali B 2010 *J. Chem. Crystallogr.* **40** 486-90
[19] Hamchaoui F, Alonzo V, Roisnel T, Rebbah H and Le Fur E 2009 *Acta Crystallogr.* **C65** i33–i35
[20] Natarajan S and Mandal S 2008 *Angew. Chem. Int. Ed.* **47** 4798-828
[21] Zhou W 2011 PhD Thesis *Proton-conducting solid acid electrolytes based upon MH(PO₃H) (M = Li⁺, Na⁺, K⁺, Rb⁺, Cs⁺, NH₄⁺* (University of Twente: the Netherlands)
[22] Oxford Diffraction 2009 CrysAlisPro CCD and CrisAlisPro RED. Oxford Diffraction Ltd, Yarnton, Oxfordshire, England

[23] Palatinus L, Steurer W and Chapuis G 2007 *J. Appl. Crystallogr.* **40** 456-62
[24] Petricek V, Dusek M and Palatinus L 2006 *Jana2006 The crystallographic computing system* Institute of Physics Praha Czech Republic
[25] ICSD, FIZ, Hermann von Helmholtz Platz 1, 76344 Eggenstein-Leopoldshafen, Germany Email: crysdata@fiz-karlsruhe.de. CCDC deposition numbers are respectively 885387 for $Cs_2[Co(H_2O)_6]_3(HPO_3)_4$ and 885388 for $Cs_2[Ni(H_2O)_6]_3(HPO_3)_4$
[26] Ouarsal R, Essehli R, Lachkar M, Zenkouar M, Dusek M, Fejfarova K and El Bali B 2004 *Acta Crystallogr.* **E60** i66-i68
[27] Ouarsal R, El Bali B, Lachkar M, Dusek M and Fejfarova K 2005 *Acta Crystallogr.* **E61** i168-i170
[28] Kratochvil B, Podlahovă J, Habibpur S, Petrícek V and Malý K 1982 *Acta Crystallogr.* **B38** 2436-38
[29] Ouarsal R, 2006 PhD Thesis *Synthèse, détermination structurale et étude spectroscopique vibrationnelle de nouveaux matériaux phosphites simples et mixtes d'éléments alcalins, alcalino-terreux et de transition 3d* (University Sidi Mohammed Ben Abdellah Fez Morocco)
[30] Ouarsal R, El Bali B, Lachkar M, Dusek M and Fejfarova K 2005 *Acta Crystallogr.* **E61** i171-i173
[31] Maalej W, Vilminot S, Elaoud Z, Mhiri T and Kurmoo M 2010 *J. Solid State Chem.* **183** 2650-55
[32] Wang X J, Zhang J H, Song J L, Kong F and Mao J G 2013 *CrystEngComm* **15** 2519-26
[33] Kosterina E V, Troyanov S I, Aslanov L A and Kemnitz E 2001 *Russ. J. Coord. Chem.* **27** 527-36
[34] Li G, Xing Y and Song S 2008 *J. Solid State Chem.* **181** 943-49
[35] Novak A 1974 *Structure and Bonding* **18** 177-216
[36] Gilli G and Gilli P 2000 *J. Mol. Struct.* **552** (1-3) 1-15

The effect of microwave and conventional heating on a modified sol-gel derived biphasic calcium phosphate

S Herradi[1,*], S Bouhazma[1], M Khaldi[1], A El Hachadi[2], B El Bali[1] and M Lachkar[1]

[1]Engineering Laboratory of Organometallic and Molecular Materials, Faculty of Sciences, University Sidi Mohamed Ben Abdellah, Po. Box 1796 (Atlas), 30000 Fez, Morocco
[2]Laboratory of Process, Signals, Industrial Systems, Computing, High School of Technology, University Cadi Ayyad, Po. Box 549, Safi, Morocco

* Email: Smaiel_her@hotmail.fr

Abstract. A facile sol-gel method was used to synthesize either hydroxyapatite (HA) or beta-tricalcium phosphate (β-TCP) as the major phase. Herein, we report, on the one hand, the effect of a very low maturation temperature on the final powder composition after drying step, and on the other hand, we compare the effect of calcination of this powder by microwave or electric furnace. It was found that microwave heating has led to the formation of hydroxyapatite phase upon 180°C for 20 minutes, however, XRD patterns show that the powder becomes less crystallized upon 220°C and amorphous upon 230°C. In contrast, furnace heating at 600°C and 700°C converts the as-synthesized powder to β-TCP as the major phase together with HA as the minor phase. This work shows the possibility to obtain the as-prepared BCP at much lower maturation temperature; it also gives an insight into the role, of either microwave or conventional heating, in controlling the ratio between HA and β-TCP in the sintered powder.

1. Introduction

Calcium phosphates (CaP) based materials are widely used in geology, biology, chemistry, dentistry and medicine [1]. Their forms are mainly described by (Ca/P) molar ratio [2, 3]. The most prominently investigated CaP are hydroxyapatite (HAp: $Ca_{10}(PO_4)_6(OH)_2$), tricalcium phosphate (β-TCP: $Ca_3(PO_4)_2$) and biphasic calcium phosphate HAp/β-TCP mixture [4]. HAp constitutes the most prevailing phase of both mammals bones (60 wt.%) and teeth (90 wt.%) [5], hence it has been used as a substitute material for teeth and bones owing to its excellent biocompatibility, osteoconductivity and stability at physiological environment [6, 7]. Owing to these advantages, researchers have synthesized HAp by various processes, including hydrothermal [8], sol-gel [9], sonochemical [10], precipitation [11], microemulsion [12], microwave [13], mechanochemical [14] and surfactant-assisted approach [15]. Among the aforementioned methods, sol-gel processes provide several advantages such as high product purity, homogeneous composition, cost-effectiveness and low synthesis temperature [16, 17]. The synthesis of HAp by sol-gel method requires that Ca/P ratio is exactly equal to 1.67 [18], however, depending on pH conditions, aging time, and temperature of the thermal treatment, variable quantities of β-TCP can be formed [19] which is beneficial since biphasic calcium phosphate (BCP), consisting of a mixture of HA and β-TCP, is considered to be more bioactive [20].

For biomaterial applications, the control of morphology and the fabrication of hydroxyapatite in much shorter time are of great interest [21]. Interestingly, in contrast to conventional heating microwave heating has many advantages such as fast synthesis, rapid volumetric heating, low energy consumption, phase purity and phase selectivity, narrow particle size distribution and facile morphology control [22].

In this work, we report the effect of very low maturation temperature on the composition of the prepared powder and we investigate the behavior of this powder under microwave or electric furnace heating. To our knowledge, this is the first work that deals with the effect of very low sol-gel medium temperature combined with a comparative study between microwave and furnace heating of the as-synthesized powder.

2. Materials and methods

2.1. Synthesis of powder

In the sol-gel process, ammonium dihydrogen phosphate ($NH_4H_2PO_4$) and calcium nitrate tetrahydrate ($Ca(NO_3)_2.4H_2O$) were selected as P and Ca precursors, respectively. Firstly, calcium nitrate (1.67 M) was dissolved in ethanol and the pH was adjusted to 10 using ammonia solution. In parallel, di-hydrogen phosphate (1M) was dissolved in deionized water and the pH was adjusted to 10. Afterward, the second solution was added dropwise to the first one. After 30 min of stirring, the mixture was matured for 48h at 4°C in the refrigerator. The gel obtained after aging process was filtered and washed repeatedly with deionized water and acetone. The gel was dried at 80°C for 12h in the oven and then was crushed using mortar and pestle. The synthesized powders are then subjected to heat treatment in an electric furnace and a domestic microwave oven for 4h and 20 min and at different temperatures viz. 600,700°C and 180, 220, 230°C respectively. In addition, pH behavior of the solution immediately after 30 min of agitation and after 48 hours of maturation at 4°C was also investigated. The flow chart of the used procedure is given in figure 1.

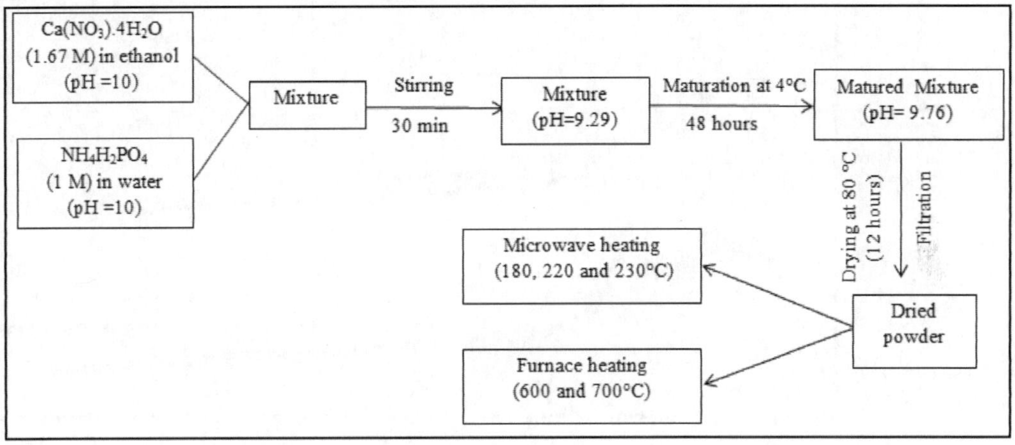

Figure 1. Flow chart of the used sol-gel process.

2.2. Characterizations

To investigate the composition and the crystallinity of the as-prepared and calcined powders, the X-ray diffraction (XRD) analysis was performed at a Discover model equipped with a monochromatized CuKα radiation (λ= 1.5418 Å) in the 2θ range (10-80°) and a step size of 0.02° and a step duration of 2s. Fourier transform infrared spectroscopy (FTIR: Bruker VERTEX 70 spectrometer) was used to analyze the functional group of the obtained powders. The spectra were recorded from 400 to 4000 cm^{-1} at 4 cm^{-1} resolution.

3. Results and discussion

The XRD patterns of the furnace and microwave heated powders are shown in figure 2 and figure 3 respectively. The pattern of the as-prepared sample shows that it is composed of hydroxyapatite and β-TCP [16, 23]. Moreover, it is clear that this sample contains amorphous phase since a few reticular planes have diffracted. Moghimian *et al* [24] reported the formation of hydroxyapatite along with calcium phosphate hydrate ($Ca_3(PO_4)_2.nH_2O$) and CaP amorphous complex when the medium temperature is 5°C, they used $Ca(OH)_2$ and H_3PO_4 as calcium and phosphorus sources. In contrast, in our as-prepared sample, no calcium phosphate hydrate has been detected. Interestingly, as it is shown in figure 2, when the as-prepared powder is calcined in an electric furnace at 600°C or 700°C it converts to BCP where β-TCP is the dominant phase [23, 25]. It is well reported that BCP can be prepared by sintering calcium deficient apatite above 900°C [26, 27]. However, our BCP could be prepared even at 600°C and that would be beneficial for improving biological responses since it has been proven that the sol-gel derived powder dimensions are in the nano-range even at high temperature [16] which mimic the nanostructure of inorganic mineral component of the natural bone matrix [28, 29].

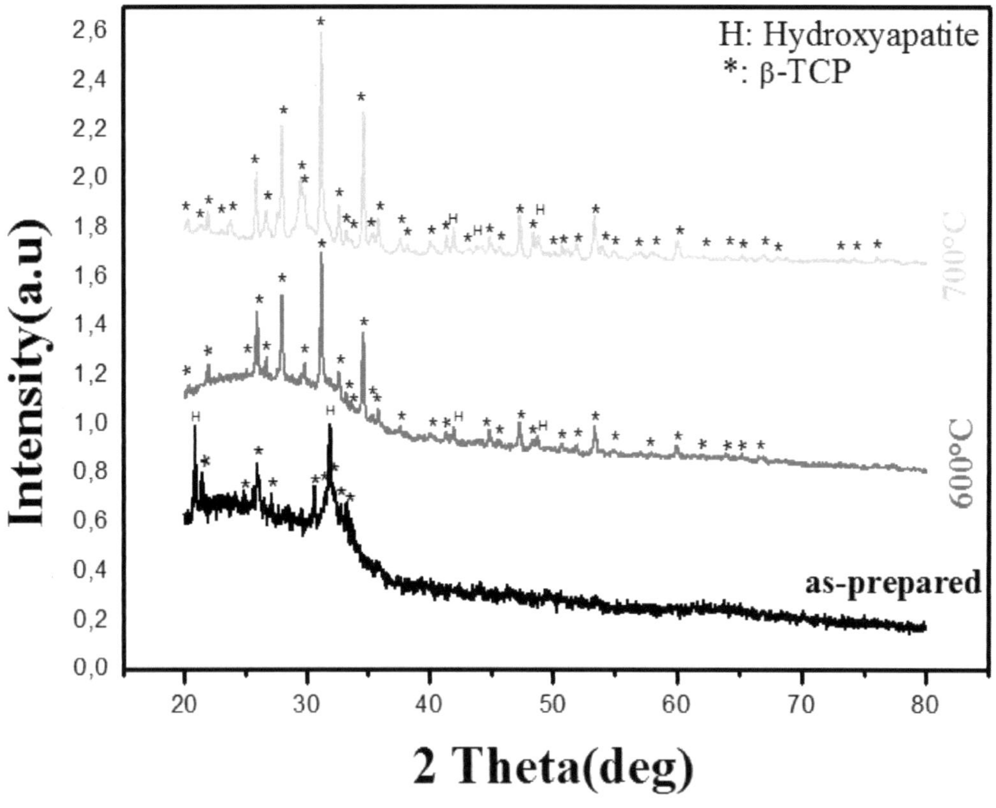

Figure 2. XRD patterns of as dried and furnace heated powders.

Conversely, as it is depicted in figure 3, microwave heated powder shows another behavior. Firstly, when heated at 180°C for 20 min the as-prepared powder becomes predominantly composed of HA phase with only one peak belonging to β-TCP [30, 31]. Secondly, at 220°C, the pattern of the as-prepared powder shows broad and weak peaks ascribed to HA phase which indicates either the nanometric nature or a low crystallinity of the powder. However, the amorphous state proved by XRD at 230°C substantiates the hypothesis that the pattern seen for 220°C is due to low crystallinity rather than the nanometric nature. Interestingly, the transformation from crystalline to non-crystalline phase has been reported for ferrites materials by Roy *et al* [32]; they found that microwave magnetic field can be used to decrystallize bulk solid matter by affecting the energy balance that determines the crystal's structure. They demonstrate also that the participation of the electrons responsible for unbalanced magnetic moments in bond formation, lead to the loss of the crystalline structure. On the basis of a thorough search in the literature, there isn't any work showing this phenomenon for hydroxyapatite. In our case, PO_4^{3-} ions possess high polarizability [33] which makes our powder absorbing microwave quickly thus enhancing the diffusion and mobility of ions which finally affect the energy balance of our crystallites and gives rise to an amorphous phase upon 230°C. In a previous work and at the same temperature [34] we have reported the formation of pure well-crystallized hydroxyapatite powder, which has been synthesized by the same procedure except for the maturation temperature, which was 25°C, and the as-prepared powder was amorphous. Based on the current results and on our previous work, we demonstrate the great effect of the low maturation temperature on the behavior of the as-synthesized powder.

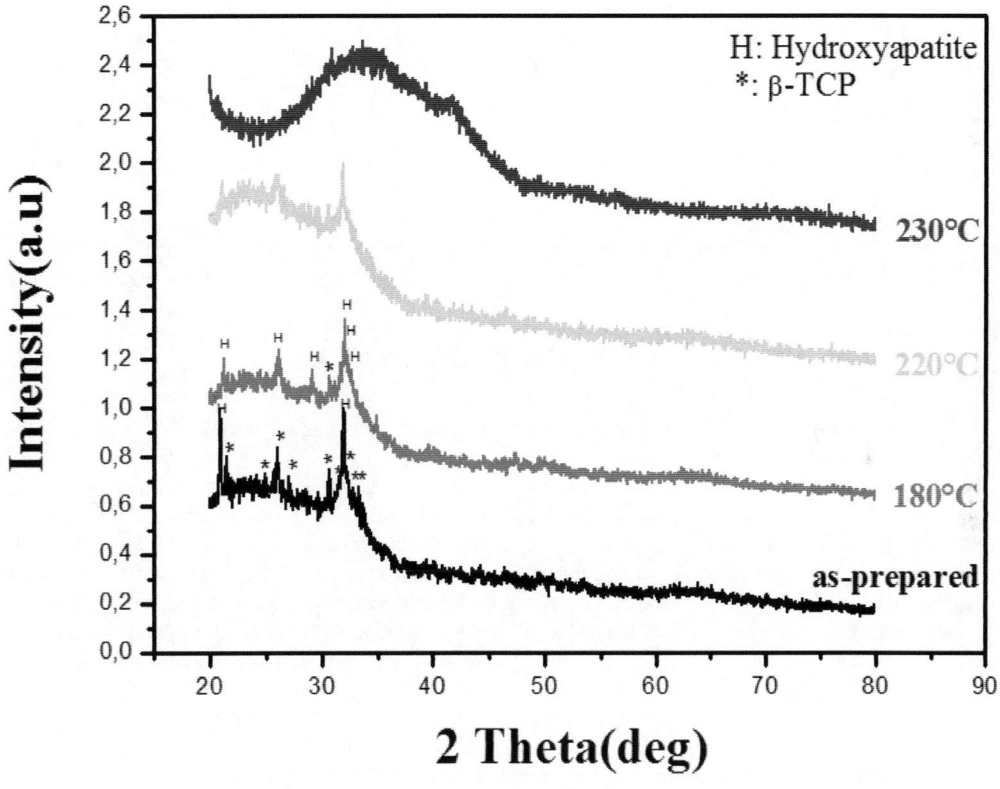

Figure 3. XRD pattern of as dried and microwave heated powders.

The IR spectra of the furnace and microwave heated powders are presented in figure 4 and figure 5 respectively. In the as-prepared sample spectrum, there are bands at 3207 and 1637 cm^{-1} ascribed respectively to the stretching mode and the bending modes of the hydroxyl group in adsorbed water, the bands at 1436 and 877 cm^{-1} are related to ν_2 bending and ν_3 stretching modes of the CO_3^{2-} group in B-type carbonated hydroxyapatite respectively [35]. The bands situated at 988 and 544 cm^{-1} are assigned respectively to a non-degenerate symmetric stretching mode, ν_1, and the doubly degenerate bending mode, v_2, of the P-O bond in phosphate group. The band located at 747 cm^{-1} is attributed to v_4 in-plane deformation bending mode of CO_3^{2-} (O–C–O bond) [36].

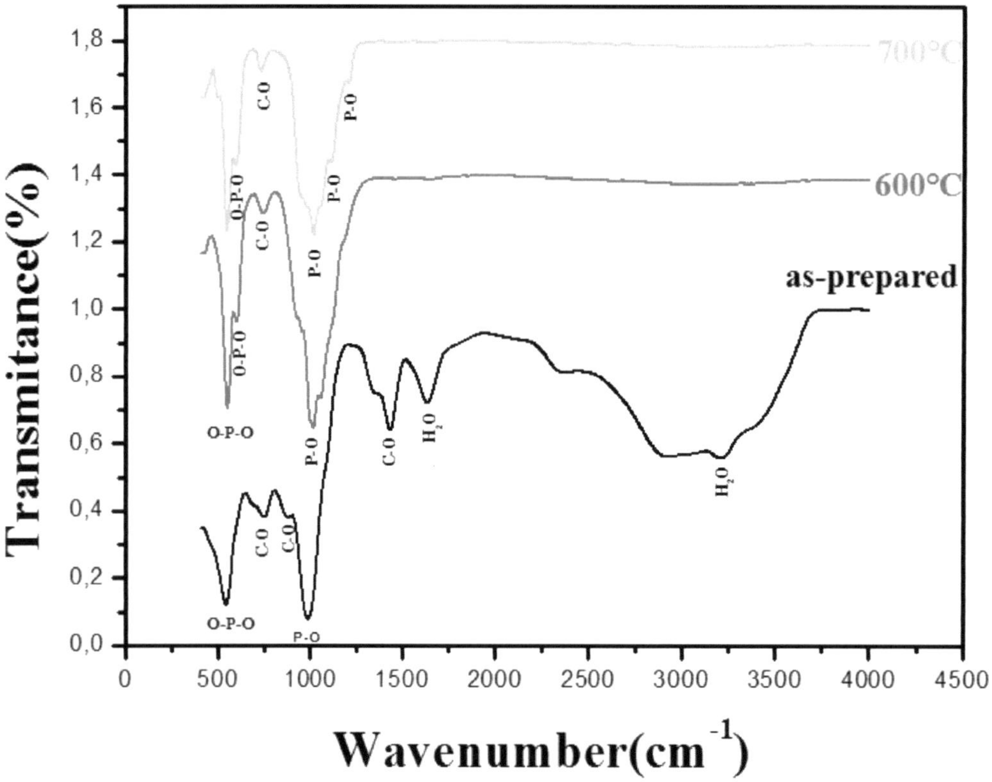

Figure 4. FTIR spectra of as-prepared and furnace heated powders.

In contrast, the spectrum of the furnace heated powder is quite different. In fact, as it can be seen in figure 4, for the powder treated at 600°C new bands has appeared at 551 and 603 cm^{-1} arise from υ_4-triply degenerate O–P–O bending in β-TCP. Moreover, the band situated at 1018 cm^{-1} is related to triple degenerated (υ_3) asymmetric stretching of phosphate groups in β-TCP [24]. Another point of particular interest is the absence of the bands related to the adsorbed water and the bands of carbonates situated at 1436 and 877 cm^{-1} which suggests a complete removal of water and a partial removal of carbonate. Furthermore, the band which remains at 742 cm^{-1} indicates that carbonated hydroxyapatite barely exists. For the sample treated at 700°C two others new bands appear in the spectrum, namely at 1117 and 1208 cm^{-1} and are due to υ_3-triply degenerate asymmetric P–O stretching in β-TCP [37].

Conversely, the spectrum of the microwave heated powder at 180°C show the characteristic bands of hydroxyapatite [3, 28, 36]. The bands located at 566 and 1032 cm^{-1} are assigned respectively to the bending vibrations (υ_4) and the asymmetric stretching vibrations (υ_3) of the PO_4^{3-} ions. The bands located at 1434, 1350, 888 and 768 cm^{-1} are attributed respectively to υ_2 bending, υ_3 stretching and υ_4 in-plane deformation bending mode of the CO_3^{2-} group in B-type carbonated hydroxyapatite [35]. In addition to the bands of adsorbed water, there is a band situated at 3393 cm^{-1} is attributed to stretching mode of crystalline OH. While the same bands are observed for the spectra of heated sample at 220°C too, the spectrum of the heated sample at 230°C show characteristic bands ascribed to amorphous calcium phosphate at 556, 1026 cm^{-1} [24]; the existence of the characteristic bands of B-type carbonate hydroxyapatite suggests that it is an amorphous carbonated hydroxyapatite.

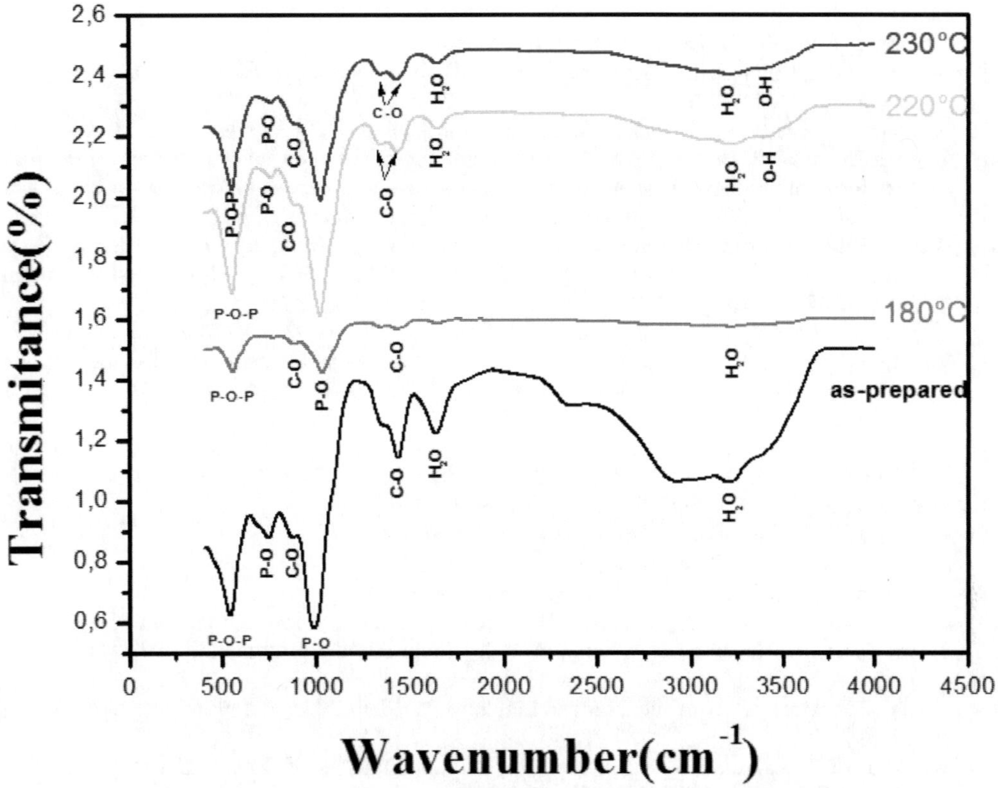

Figure 5. FTIR spectra of as-prepared and microwave heated powders.

The pH behavior explains clearly what has happened in the medium. Directly after adding the first drops of ammonium dihydrogen phosphate solution (pH=10) into the calcium nitrate solution (pH=10) we have observed that the pH was decreasing continuously and immediately after 30 min of stirring the pH has dropped to 9.29 which means that there was a consumption of OH^- owing to the nucleation of hydroxyapatite. However, after 48 hours of maturation at 4°C, we have noticed that the pH has increased to 9.76 due to the release of hydroxyl ions from the formed hydroxyapatite nuclei thus explaining the biphasic nature of the as-prepared powder.

4. Conclusion

The effect of low maturation temperature and microwave or furnace heating was studied. It was found that low maturation temperature has led to biphasic calcium phosphate containing hydroxyapatite and β-TCP. Furthermore, microwave heating has led to hydroxyapatite upon heating at 180°C and to amorphous phase upon heating at 230°C. Conversely, furnace heating has led to BCP where β-TCP is the major phase.

Acknowledgements
The authors would like to acknowledge the support and technical assistance of Interface Regional University Center (University Sidi Mohammed Ben Abdellah, Fez), and National Center for Scientific and Technical Research (CNRST-Rabat).

References

[1] Dorozhkin S V 2007 *J. Mater. Sci.* **42** 1061–95
[2] Destainville A, Champion E, Bernache-Assollant D and Laborde E 2003 *Mater. Chem. Phys.* **80** 269–77
[3] Raynaud S, Champion E, Bernache-Assollant D and Thomas P 2002 *Biomaterials* **23** 1065–72
[4] Berzina-Cimdina L Borodajenko N 2012 Research of calcium phosphates using Fourier transform infrared spectroscopy *Infrared Spectroscopy Materials Science Engineering and Technology* ed T. Theophanides (Rijeka Croatia: Intech) chapter 6 pp 123–48
[5] Kaur J 2010 *Properties of biologically relevant nanocomposites: effects of calcium phosphate nanoparticle attributes and biodegradable polymer morphology* ed Proquest (Georgia, Institute of Technology: UMI Dissertation Publishing) p 16
[6] Yoruça H, Karakaşb A, Koyunc A and Yildiz T 2012 *Acta. Phys. Pol.* A **121** 233–5
[7] De Melo Costa A C F, De Lima M G, De Almeida Lima L H M, Cordeiro V V, De Souto Viana K M, De Souza C V and De Lucena Lira H 2009 *Revista eletrônica de Materiais e Processos* **4** 29–38
[8] Vázquez-Hernándeza F, Mendoza-Barrera C, Altuzar M, Meléndez-Lira M A, Santana-Aranda M, Olvera M and De La Olvera L 2010 *Mater. Sci. Eng.* B **174** 290–5
[9] Fathi M H and Hanifi A 2007 *Mater. Lett.* **61** 3978–83
[10] Giardina M A and Fanovich M A 2010 *Ceram. Int.* **36** 1961–9
[11] Li Z, Wang P and Wu Z 2005 *J. Mater. Sci.* **40** 6589–91
[12] Bose S and Saha S K 2003 *Chem. Mater.* **15** 4464–649
[13] Mishra V K, Srivastava S K, Asthana B P and Kumar D J 2012 *J. Am. Ceram. Soc.* **95** 2709–15
[14] Nasiri-Tabrizi B, Honarmandi P, Ebrahimi-Kahrizsangi R and Honarmandi P 2009 *Mater. Lett.* **63** 543–6
[15] Coelho J M, Moreira J A, Almeida A and Monteiro F J 2010 *J. Mater. Sci.: Mater. Med.* **21** 2543–9
[16] Kuriakose T A, Kalkura S N, Palanichamy M, Arivuoli D, Dierks K, Bocelli G and Betzel C 2004 *J. Cryst. Growth* **263** 517–23
[17] Choi A H and Ben-Nissan B 2007 *Nanomedicine* **2** 51–61
[18] Liu D, Troczynski T and Tseng W J 2001 *Biomaterials* **22** 1721–30
[19] De Aza P N, Guitian F, Santos C, De Aza S, Cusco R and Artus L 1997 *Chem. Mater.* **9** 916–22
[20] Daculsi G, LeGeros R Z, Nery E, Lynch K and Kerebel B 1989 *Biomed. Mater. Res.* A **23** 883–94
[21] Lerner E, Sarig S and Azoury R 1991 *J. Mater. Sci. Mater. Med.* **2** 138–41
[22] Zhu Y J and Chen F 2014 *Chem. Rev.* **114** 6462–555
[23] Wang J and Shaw L L 2009 *Biomaterials* **30** 6565–72
[24] Moghimian P, Najafi A, Afshar S and Javadpour J 2012 *Adv. Powder Technol.* **23** 744–51
[25] Rangavittal N, Landa-Cánovas A R, González-Calbet J M and Vallet-Regí M 2000 *J. Biomed. Mater. Res.* A **51** 660–8
[26] LeGeros R Z and Daculsi G 1990 In vivo transformation of biphasic calcium phosphate ceramics: ultrastructural and physico-chemical characterization *Handbook of Bioactive Ceramics* ed T. Yamamuro, L. L. Hench and J. Wilson (Boca Raton: CRC Press) pp 17–28
[27] Bouler J M, LeGeros R Z and Daculsi G 2000 *J. Biomed. Mater. Res.* **51** 680–4
[28] Poinern G J E, Brundavanam R. L. X. T., Le X T, Djordjevic S, Prokic M and Fawcett D 2011 *Int. J. Nanomed.* **6** 2083–95
[29] Kalita S J, Bhardwaj A and Bhatt H A 2007 *Mater. Sci. Eng.* C **27** 441–9
[30] Ma T, Xia Z and Liao L 2011 *Appl. Surf. Sci.* **257** 4384–8
[31] Zyman Z, Goncharenko A, Rokhmistrov D and Epple M 2011 *Material. wiss. Werkstofftech.* **42** 154–7
[32] Roy R, Peelamedu R, Grimes C, Cheng J and Agrawal D 2002 *J. Mater. Res.* **17** 3008–11
[33] Cao J M, Feng J, Deng S G, Chang X, Wang J, Liu J S, Lu P, Lu H X, Zheng M B, Zhang F and Tao, J 2005 *J. Mater. Sci.* **40** 6311–13

IOP Conf. Series: Journal of Physics: Conf. Series **984** (2018) 012016 doi:10.1088/1742-6596/984/1/012016

[34] Herradi S, El Bali B, Khaldi M and Lachkar M 2017 *IOP Conf. Series: Materials Science and Engineering* **186** 012023
[35] Ślósarczyk A, Paszkiewicz Z and Paluszkiewicz C 2005 *J. Mol. Struct.* **744** 657–61
[36] Garskaite E, Gross K A, Yang SW, Yang T C K, Yang J C and Kareiva A 2014 *Cryst. Eng. Comm.* **16** 3950–9
[37] Nandha Kumar P, Ferreira J M D F and Kannan S 2017 *Inorg. Chem.* **56** 1289–99